NEW WAVE MANUFACTURING STRATEGIES

Human Resource Management Series: Foreword

The two integrating themes of this series are organizational change, and the strategic role of the human resource function.

The 1980s witnessed a fundamental shift in thinking with respect to the organizational role of the personnel function. That shift is typically reflected in the change of title – to human resource management – and revolves around the notion that effective human resource management is a critical dimension of an organization's competitive advantage. Personnel or human resource management is thus now more widely accepted as a strategic business function, in contrast with the traditional image of a routine administrative operation concerned with hiring, training, paying and terminating.

The range of issues with which personnel managers must now deal has widened considerably, as has the complexity and significance of those issues. Conventional texts in this subject area typically have the advantage of comprehensiveness, by offering a broad overview of the function, its responsibilities, and key trends. Such coverage is always bought at the expense of depth. The aim of this series, therefore, is not to replace traditional personnel or human resource texts, but to complement those works by offering in-depth, informed and accessible treatments of important and topical themes, written by specialists in those areas and supported by systematic research.

The series is thus based on a commitment to contemporary changes in the human resource function, and to the direction of those changes. This has involved a steady shift in management attention towards improved employee welfare and rights, genuinely equal opportunity, wider employee involvement in organization management and ownership, changing the nature of work and organization structures through business 'reengineering', and towards personal skills growth and development at all organizational levels. This series documents and explains these trends and developments, indicating the progress that has been achieved, and aims to contribute to best management practice through fresh empirical evidence and practical example.

David Buchanan
Series Editor
Loughborough University Business School

NEW WAVE MANUFACTURING STRATEGIES: ORGANIZATIONAL AND HUMAN RESOURCE MANAGEMENT DIMENSIONS

Edited by
John Storey
Loughborough University Business School

Paul Chapman Publishing Ltd
144 Liverpool Road
London
N1 1LA

British Library Cataloguing in Publication Data

New Wave Manufacturing Strategies
I. Storey, John
658.5

ISBN 1 85396 180 9

Typeset by Inforum, Rowlands Castle, Hants
Printed and bound by Athenaeum Press, Newcastle upon Tyne

B C D E F G H 9 8 7

CONTENTS

CONTRIBUTORS

Paul Adler, Professor of Engineering Management, School of Business, University of Southern California, Los Angeles, USA.

Harry Boer, Professor of Production and Operations Management, School of Management, University of Twente, The Netherlands.

David Buchanan, Professor of Human Resource Management and Director of Loughborough University Business School.

Patrick Dawson, Senior Lecturer in Organisation Studies, Department of Commerce, The University of Adelaide, Australia.

Arthur Francis, Professor of Strategic Management, Glasgow University Business School, University of Glasgow.

Alan Harrison, Lecturer in Production and Operations Management, Warwick Business School, University of Warwick.

Malcolm Hill, Professor of Russian and East European Industrial Studies, Loughborough University Business School.

Nicholas Kinnie, Senior Lecturer, School of Management, University of Bath.

Alan Spreadbury, MRPII Project Manager, Dista Products, UK.

Roy Staughton, Lecturer in Production and Operations Management, University of Bath.

John Storey, Senior Lecturer and Director of the Human Resource and Change Management Research Unit, Loughborough University Business School.

Joe Tidd, Lecturer, The Management School, Imperial College, London.

PREFACE

The vital importance of a thriving manufacturing sector for the viability of the national economy as a whole is now widely recognized. To survive, under the new conditions in which modern manufacturing occurs means, however, having to compete on new terms. International companies have adopted new principles of manufacturing which have taken competition on to new planes. A whole array of initiatives have been introduced over the past decade or so (FMS, JIT, TQM, CIM and so on). It is the far-reaching significance of these new approaches – collectively labelled here 'New Wave Manufacturing' – which prompts the writing of this book,

A very considerable body of research evidence as well as reports from practitioners' own experiences make one point time and time again: it is the organizational and Human Resource Management aspects of the new strategies which are critical to their success or failure. Few studies to date have, however, elaborated the details of this observation. Many questions remain unanswered but one predominating question provides the driving force for this book – to what extent do the various new approaches, be they JIT, CIM, MRPII or TQM – require a similar set of organizational and Human Resource policies and practices? Or is it the case that each 'technique' has its own particular HR and organizational requirements?

To answer these sorts of questions requires a very special approach. The answers are not to be found easily in the currently available textbooks or even in research articles. This book is distinctive in that it attends to the problem through the joint eyes of production and behavioural analysts. As such, the book will serve as a unique introduction to the meaning of the new developments in manufacturing methods *and*, as added value, provide an up-to-date assessment of the HR and organizational dimensions to these methods.

The chapters were commissioned from leading experts in their respective fields. They were all given the same specifications – to produce chapters which would give clear, accessible, up-to-date reviews of their topic areas and to add, where appropriate, the latest findings from their own research, consultancy and practice. The authors are drawn from Europe, the USA and Australia as well as from the UK.

Another distinctive feature of this book is that it treats the problem of New Wave Manufacturing in a *holistic* way. That is, unlike most other reviews, this book does not confine its attention simply to shopfloor issues. The whole production cycle including, for example, design, product planning, engineering strategy, technology and implementation is brought into the analytical frame.

New Wave Manufacturing Strategies accordingly has a vision which goes beyond the so far predominant 'new technology'/Advanced Manufacturing Technology discussions (for examples of these see Clark (ed.) (1993); Wall *et al.* (eds.) (1987); Warner *et al.* (eds.) (1990).

We have three main audiences in mind. In no particular order these are: first, practising managers who want both an introduction to the new methods and an authoritative state-of-the-art review of these developments; second, students of manufacturing engineering, production management, and a whole host of other engineering students – chemical engineering, mechanical engineering and others who are preparing for careers in manufacturing; and third, students from the social sciences, business and management who need to examine organizational and behavioural issues in manufacturing contexts. This intermingling of audiences and perspectives is of extreme significance. Only by gaining a greater understanding of technology and production methods will social scientists be able adequately to address the political, ethical and social issues which interest them. Likewise, only by attending to the behavioural issues will engineers and production specialists find themselves capable of implementing the systems which they have helped formulate.

Acknowledgements are due to those manufacturing managers from many sectors who have co-operated with our research efforts over many years and who encouraged me to compile this book. And thanks to Lynne Atkinson at Loughborough for excellent secretarial support.

John Storey
November 1993

REFERENCES

Clark, J. (ed.) (1993) *Human Resource Management and Technical Change*, Sage, London.

Wall, T. D., Clegg, C. W. and Kemp, N. J. (eds.) (1987) *The Human Side of Manufacturing Technology*, John Wiley, Chichester.

Warner, M., Wobbe, W. and Brodner, P. (eds.) (1990) *New Technology and Manufacturing Management: Strategic Choices for Flexible Production Systems*, John Wiley, Chichester.

ABBREVIATIONS

AMT	Advanced Manufacturing Technology
BPR	Business Process Re-engineering
CAD	Computer-Aided Design
CAE	Computer-Aided Engineering
CAM	Computer-Aided Manufacture
CAPM	Computer-Aided Production Management
CAPP	Computer-Aided Process Planning
CIM	Computer-Integrated Manufacturing
CM	Cellular Manufacturing
CNC	Computer Numerical Control
CRP	Capacity Requirements Plan
DFM	Design for Manufacture
DNC	Direct Numerical Control
FMC	Flexible Manufacturing Cells
FMM	Flexible Manufacturing Module
FMS	Flexible Manufacturing System
GMP	Good Manufacturing Practice
GT	Group Technology
HR	Human Resource(s)
HRM	Human Resource Management
IT	Information Technology
JDS	Job Diagnostic Survey
JIT	Just-in-Time
LAN	Local Area Network
LP	Lean Production
MBO	Management by Objectives
MPS	Master Production Schedule
MRP	Materials Requirement Planning
MRPII	Manufacturing Resource Planning

MSE	Manufacturing Systems Engineering
NC	Numerical Control
NPD	New Product Development
NWM	New Wave Manufacturing
OD	Organization Development
OEM	Original Equipment Manufacturer
OPT	Optimized Production Technology
QFD	Quality Function Deployment
QWL	Quality of Working Life
R&D	Research and Development
S&OP	Sales and Operations Planning
SPC	Statistical Process Control
SUR	Set-up Reduction
TPM	Total Productive Maintenance
TQ	Total Quality
TQC	Total Quality Control
TQM	Total Quality Management
TQS	Total Quality Service
WIP	Work in Progress
WTL	Work-to List

LIST OF FIGURES

1

NEW WAVE MANUFACTURING STRATEGIES: AN INTRODUCTION

John Storey

In response to radical changes in the nature and degree of international competition, many companies have begun to embark upon equally radical new ways of manufacturing. These 'new ways' have been given a bewildering array of labels: Lean Production (LP), Just-in-Time (JIT), the Integrated Factory, World Class Manufacturing, Total Quality Management (TQM), Computer-Integrated Manufacturing (CIM), Cellular Manufacturing (CM) and Flexible Manufacturing Systems (FMS) – to name just a few of the most significant types of innovation. While it is true that speculation (for example, on the theme of the fully automated 'un-manned factory') has often run way ahead of actual practice it is also the case that, in many ways, research-based knowledge has also lagged seriously behind actual developments and installations. Nowhere is this statement more true than with regard to the organizational and Human Resource Management aspects of these new forms of manufacturing.

There is a fragmented, inconclusive, partial and often contradictory set of literatures associated with the phenomenon. Much of the discussion attends to the issue as if it were a continuation of the problem of 'new technology' (Warner, Wobbe and Brodner, 1990; Wall, Clegg and Kemp, 1987). But in fact, many elements of the 'new manufacturing' are only partially connected with technology as conventionally understood. Of equal importance are the so-called 'softer' elements including, for example, internal customer supply chains, new ways of working, cellular factory lay-out, and so on. One of the most wide-ranging claims is that the era of mass-production has reached the end of its life cycle and is being superseded by some new form such as 'flexible specialization' (Piore and Sabel, 1984) or 'Lean Production' (Womack, Jones and Roos, 1990), 'Toyotism' (Imai, 1986), or 'post-Fordism' (Roobeck, 1987; Kenney and Florida, 1988, 1993). Whatever the appropriate terminology, the issues themselves have triggered debate in a host of countries, Germany and the USA in particular. For example, it is estimated that there have been more than fifty conferences in Germany on the subject of

'Lean Production' alone in the past couple of years (Cooke, 1993: 77). Another literature-set addresses the topic from the viewpoint of Production and Operations Management (Hill, 1986, 1991; Voss, 1986; Voss and Robinson, 1987). An especially notable contribution in this category is the work of Gerwin and Kolodny (1992). This wide-ranging analysis incorporates the three elements of strategy, organization and innovation in a coherent framework.

A further alternative has been to view recent manufacturing innovations in the West as an emulation of the successful Japanese model – hence the now popular label 'Japanization' to describe the phenomenon (Turnbull, 1986, 1988; Oliver and Wilkinson, 1992; Bratton, 1992). Schonberger (1982, 1986) is the original source of much of this discussion. He refers to the 'upheaval' in manufacturing and draws attention to Japanese manufacturing concepts and techniques which are '180 degrees out of phase with our own' (1986: ix). These types of 'World Class Manufacturing' methods can also be found, however, adopted in Western companies. The characteristic features are:

> stock-rooms emptied and converted into manufacturing; storage racks, automated handling systems and conveyers dismantled; fork lift trucks eliminated; complicated and costly computer systems replaced by manual charts and blackboards with data entered and interpreted by operators; existing machines upgraded to process capability and moved into cells that make and feed parts forward just in time for the next machine to use them; reduced numbers of inspectors, suppliers, and part numbers; whole layers of management eliminated.
>
> (Schonberger, 1986: ix)

In similar apocalyptic fashion, Womack *et al.* (1990) chart the emergence of 'Lean Production'. This step-change in manufacturing methods, they say, 'combines the advantages of mass production and craft production'. It is based on teams of multiskilled workers and flexible automated machines. It is 'lean' in that it uses less human effort, half the manufacturing space, half the investment in tools, half the engineering hours to produce a new product and has less inventory. The new paradigm 'sets its sights' on zero defects and zero inventories. It also 'changes how people work'. For most people, according to Womack *et al.* (1990: 14), this will mean more challenging jobs, greater responsibility and teamworking rather than narrow professional careers. In sum, Lean Production is quite simply 'a new way of making things' (1990: 47).

Other, alternative, approaches to understanding the new manufacturing methods include studies in the industrial sociology and labour process traditions (Braverman, 1974; Williams *et al.*, 1992); studies which see the developments as illustrative of a move to 'post-modernism' (Clegg, 1992); and studies which draw on the philosophical work of Foucault in order to highlight the pervasive control elements in the new methods (Sewell and Wilkinson, 1991).

When compared to the above sets of literature, the distinctive focus and character of this book is twofold. First, there is a conscious melding of the production management perspective with the behavioural perspective. Accordingly, the various chapters are written by expert contributors from both these traditions and editorial interventions have been made during the production of the chapters to help ensure that aspects of each perspective inform every chapter. Second, and very much related to the first point, the central theme of the book is the analysis of the *organizational and Human Resource Management dimensions* of the new manufacturing. In order to explore these in detail, close analysis is made of the production management and technological characteristics of the new manufacturing forms. In this way, the book should serve as a sound introduction to the production management innovations *per se* as well as adding value by offering fresh insights into the organizational and Human Resource aspects. What is particularly critical to note at this juncture is that the analysis presented in this book seeks to go way beyond the now tiresome platitude that 'the human side' is important and deserves greater attention. Of course it is and of course it does, but what still needs exploring and explaining are the actual ways in which this motherhoood statement can be borne out.

The organizational and Human Resource (HR) agenda on New Wave Manufacturing (NWM) is potentially vast. As noted above, disconnected elements of it have been variously addressed. But there is scope to put greater order and clarity upon this field. Essentially there are just two main sets of issues which concern the HR/organizational dimension in NWM. The first set relates to features deemed to be required for the actual realization of the new manufacturing techniques and technologies. This agenda tackles the social *prerequisites* for successful implementation of Advanced Manufacturing Technology. A subsidiary question is the way in which organizational and HR practices play a part in the actual implementation processes. Here, the issue is the HR aspects of the management of the processes of change.

The second set of issues relates to the HR/organizational *outcomes* and impacts of NWM methods. Under this heading can be placed the negative and positive impacts on worker skill levels, on work intensity, task variety, employee morale and job satisfaction. In essence then, we will need to look at the HR and organizational dimensions both as inputs to, and outputs from, NWM strategies.

Drawing upon the available literature as well as upon a range of new research, the chapters in this book will contribute to a deeper level of understanding on each of these areas of enquiry. In order to set the scene, this chapter will present an introductory overview so as to situate the relevant issues. There is potentially a vast range of these to consider. They will be discussed in the following three sections which constitute the main body of this chapter.

In section one an analysis is made of the *meanings and characteristics* of NWM methods. Section two explores the reasons for the adoption of these methods, the planned benefits and the *actual outcomes* as so far recorded – both positive and negative. The third section examines the *barriers* to successful implementation. This discussion embraces both the infrastructural *prerequisites* and the organizational/HR dimensions involved in the management of the change processes during implementation. The final section offers a synopsis of the ensuing chapters and highlights their particular contributions.

MEANINGS AND CHARACTERISTICS OF NEW WAVE MANUFACTURING

As we noted above, the new methods of manufacturing have been given a variety of labels. One of the most important questions to be addressed therefore is the extent to which these are merely different labels for the same basic set of phenomena or whether the different terms signify critically different approaches. The related vital question from our viewpoint is whether any of the differences in manufacturing method carry significantly different implications for Human Resource Management (HRM) or whether each of the new methods broadly requires the same set of HRM policy choices and practices.

The keywords associated with the new manufacturing management are 'flexibility', 'quality', 'teamworking', 'Just-in-Time' delivery, 'right-first-time' production, the elimination of waste and non-value-added activity, 'zero-defects' and 'continual improvement'. Less certain is the extent to which it is possible to install individual elements or whether there is a coherent 'package' which, in effect, has to be implemented as a whole.

Arguably, the separate elements are not so new. What is different is the way the NWM approach brings them all together into a new, mutually reinforcing, whole. Hence, JIT places a premium on right-first-time and Total Quality; continuous improvement requires involvement from everyone and some form of teamworking; in turn, teamworking implies a need for flexibility, while flexibility means a better trained and more competent workforce.

The problem is that large numbers of management consultants and many, if not indeed most, contributions to the literature on manufacturing, tend to concentrate on just one or other of the 'new' methods. Hence, the journals are replete with articles which extol the merits of particular approaches such as JIT, Manufacturing Resource Planning (MRPII) or Cellular Manufacture with little or no regard to how the elemental features of the method being promoted stand *vis-à-vis* other current methods. This neglect is especially irksome given that many of the characteristics supposedly inherent to many of these 'different' methods turn out to be so similar.

There are, however, a few notable exceptions to the singular-method contributions (see, for example, Drucker, 1990; Noori, 1990). Drucker's account of the 'emerging theory of manufacturing' discusses four concepts that 'together constitute a new approach to manufacturing' (Drucker, 1990: 94). The

Common abbreviations	Full title	Main characteristics
JIT	Just-in-Time	Elimination of buffer-stocks; delivery of materials and sub-assemblies just in time to be worked on
TQM	Total Quality Management	Full participation of every function and person in the organization in producing goods and services to their customers' requirements; continuous process improvement; application of relevant techniques to support the above (e.g. statistical process control)
LP	Lean Production	Teamwork; continuous improvement; zero defects; JIT; the integration of suppliers
MRP	Materials Requirement Planning	A materials control system based on the forward production plan
MRPII	Manufacturing Resource Planning	An integrated approach to production planning and control using a computer-based information system which also carries far-reaching organizational change implications
CIM	Computer-Integrated Manufacturing	A computer-based system which integrates all elements in the manufacturing process from product design to distribution

Figure 1.1 A summary of some New Wave Manufacturing methods

four concepts (Statistical Quality Control, a new accounting method, modular organizational forms, and a systems approach) although being developed and promoted separately, he says, are none the less 'synergistic'. Together, 'but only together, they tackle the conflicts that have most troubled traditional, twentieth century mass production plants' (Drucker, 1990: 102).

There is, however, a reverse logic which could be applied to concepts such as JIT, CIM, MSE and TQM. Given that each of these is itself wide-ranging and diffuse, it is possible to question their internal consistency rather than search for inter-connections. It is of course also possible to do both, that is to seek common elements between the apparently different techniques *and* to probe their internal consistencies. So, while it is possible on the one hand that the myriad of new manufacturing methods may in fact share many common characteristics it is also possible that the terms themselves suggest an internal consistency which in reality is lacking. In order to explore the meanings and characteristics further it will be found helpful to refer to Figure 1.1 which summarizes the main approaches.

It is useful to begin with a preliminary discussion of some of the main features of one of the most notable forms of the new management – JIT. This is shown as the first item in Figure 1.1. Since the 1970s the approach has been widely known in Japan as the Toyota Manufacturing System. One of the foremost writers on Japanese manufacturing has described JIT as a system designed 'to produce and deliver finished goods just in time to be sold, sub-assemblies just in time to be assembled into finished goods, and purchased materials just in time to be transformed into fabricated parts' (Schonberger, 1982: 19). JIT is a methodology which aims to achieve the virtual elimination of stock levels at all points in the production process – from materials inward, Work in Progress on the shopfloor and warehousing prior to dispatch. Not only are savings made because cash is released from lower stock levels, but, in addition, the removal of buffer-stocks helps to expose problems in the production process. JIT has accordingly been described as 'a philosophy to eliminate waste where waste is anything that adds cost but not value to a product, value being what the customer is prepared to pay for' (Lucas Engineering, 1991: 4.1).

Under JIT, production is 'pulled' through the system as and when it is required. '*Kanban*' is the simple card-and-box device which facilitates the pull type of material control. The *kanban* card informs operatives of the specifications of components to be made, their number and their next destination.

Beyond these basic features there lies a good deal of controversy. For example, many commentators refer to the lack of consensus about the true meaning of JIT. It is not just a single technique but a 'toolbox' of techniques (Graham, 1988). This point is borne out by the results of a survey of JIT programmes by Voss and Robinson (1987). They discovered that these programmes comprised a whole series of elements, some of which were utilized frequently while others were activated only rarely. For example, of those purportedly having a JIT programme, 80% said they were using labour more flexibly, 44% were using cells but only 11%, surprisingly, were using *kanban*.

In other words, the understanding of what JIT in its wider sense actually consists of is rather varied. Monden (1983), one of the most cited sources, describes it as cost reduction through the elimination of all types of waste. Others refer to it as a complicated 'philosophy' (Safayeni *et al.*, 1991; Sohal, Keller and Fouad, 1989; Westbrook, 1987).

Our own research in manufacturing organizations suggests that while some managers view it as a near total system of continuous improvement, others simply regard it as essentially the *kanban* pull system and make few other claims for it or upon it.

A full analysis of the meanings of JIT and the organizational and HRM dimensions is made in Chapter 9.

The way the initiatives such as JIT, CIM, FMS, TQ and Materials Requirement Planning (MRP) or MRPII are presented often makes it appear a foregone conclusion that they are 'alternative' and even 'competing' systems. It is common, for example, to see JIT sold as 'superior' to the computer-driven controls of MRP. But as Karmarkar (1989) among others argues, they are not, in fact, necessarily alternatives. JIT and *kanban* systems are used successfully by those companies that are also leaders in advanced computer applications and MRPII – Hewlett Packard and Toyota to give two examples. The reason is that in practice companies need both, that is shopfloor devices such as *kanban* as well as planned, computer-driven push systems such as MRPII. There is in consequence 'no need to choose between push or pull. These methods are not mutually exclusive and each has its pros and cons. The best solution is often a hybrid that uses the strengths of both approaches' (Karmarkar, 1989: 27). An example of the hybrid type is 'factory management systems', a new approach which uses computer technologies geared to shopfloor use such as bar-coding and smart cards. In essence, one is looking forward here to advances in IT which allow decentralization of scheduling decisions to cell level whilst still meeting market needs in an optimized way.

Our experience suggests that there are different stages of take-up of the new manufacturing methods. Large numbers of organizations are merely at the first stage. They have toyed with the easy bit: consultants have been engaged and they have embarked upon educational, awareness-raising and training programmes. The 'common language' of the new methods has been learned. But what happens next? Safayeni *et al.* (1991) have usefully identified four 'levels' of implementation of JIT. By extension, the same could be said to apply to new wave management approaches in general. At level one there is merely a good deal of familiarization and 'talking' about the new way. Some minor modifications may be made and there are likely to be various attempts to reclassify certain already existing initiatives and behaviours within the label. At level two, pilot projects are launched. The attempt is to try it out and make it a winning showcase without disturbing the main production system. Getting beyond this stage is often a major problem. Level three implementation is characterized by the launch of JIT in one part of the production process. At this stage JIT is vulnerable to dissipation as it is overwhelmed by the predominant system. Level four is the stage of total take-up of JIT. The organization is restructured along product lines – the manufacturing system and the support functions alike. Such organizations are few and far between. Instead of a radical shift to product-focused organizations, 'it is much easier for them to "fit" JIT into their existing structure . . . however, it is possible to argue that such a reorganization may be one of the prerequisites of successful JIT "implementation"' (Safayeni, 1991: 33).

It can be said, in summary, that despite the variety of different labels for the new methods of manufacturing there are many common elements. When added together they can fairly be claimed to constitute a new paradigm. While 'new' technology (AMT) plays a part in this, the crucial point to note is that the 'new wave' is, above all, a different set of arrangements for the *social organization of production*. The key characteristics associated with this systemic break with conventional mass Fordist production include flexibility, teamworking, continual improvement and adaptation, and integration.

INTENTIONS AND OUTCOMES

The sought-for benefits from the array of new manufacturing methods and concepts are numerous. There is, however, a good deal of overlap between the claimed benefits associated with the separate approaches. Most of the advocates of the various methods suggest that inventories and Work in

Progress will be drastically reduced, that quality will be enhanced, that components and sub-assemblies will be made 'right-first-time', that flexibility will be enhanced, productivity increased, waste of all kinds reduced or eliminated, value-added increased and, above all, competitiveness notched up a goodly number of turns.

The promised and hoped-for benefits do not of course always materialize. The outcomes are sometimes very different. In this section we review the findings on stated intentions and actual outcomes.

The *positive* benefits of JIT have been described by Monden (1983), Cheng (1988) and Schonberger (1982, 1986). These outcomes include improved relations with suppliers, higher productivity, savings because of low inventory and, crucially, the highlighting of, and subsequent elimination of, underlying problems formerly disguised by the buffer system. For example, Schonberger (1986: ix) reports that the new World Class Manufacturing methods result in 'defect rates cut tenfold and more; manufacturing lead-times cut twentyfold in a number of cases; triple the sales volume in half the plant space'. Oliver (1989) reports case studies where the outcomes of JIT included improvements in throughput, quality and lead times as well as cell members displaying a sense of 'ownership' for their part of the process.

Improved flexibility is reported by Meredith (1987) and Skinner (1985). Cost reductions are cited by Schonberger (1982) and Skinner (1985). Reductions in the requirement for indirect labour are claimed by Monden (1983) and Meredith (1987).

The most impressive sets of data on the performance outcomes of NWM methods are those which tabulate the measures of the Toyota manufacturing system with the traditional mass-assembly methods in Western car plants. For example, Womack *et al.* (1990: 81) present data of this kind in Figure 1.2.

In terms of the impact on work and workers' opinions there is extensive controversy. Womack *et al.* (1990: 101) cite United Automobile Workers' criticism of Lean Production as 'management by stress' because managers

	Western Assembly Plant (GM, Framingham)	Japanese Assembly Plant (Toyota, Takaoka)
Gross Assembly Hours per car	40.7	18.0
Adjusted Assembly Hours per car	31	16
Assembly Defects per 100 cars	130	45
Assembly Space (square feet per car per year)	8.1	4.8
Inventory of Parts (average)	2 weeks	2 hours

Notes: Gross assembly hours per car are calculated by dividing total hours of effort in the plant by the total number of cars produced.

'Adjusted assembly hours per car' incorporates the adjustments described in the Womack *et al.* text.

Source: Womack *et al.* 1990, p. 81.

Figure 1.2 Comparison of a Western Mass Production Plant and a Japanese Lean Production Plant

continually seek to eliminate 'slack' from the system. This means that unused work time, excess workers and excess inventories are targeted. Also, Lean Production is compared unfavourably with the group-based production methods at Volvo in Sweden. But Womack *et al.* seek to refute both points. They say that Lean Production offers workers the skills they need to meet the new challenges and that 'creative tension' is induced (Womack *et al.*, 1990: 102). Womack and colleagues are dismissive of the 'neocraftsmanship' of Volvo, arguing that 'simply bolting and screwing together a larger number of parts in a long cycle rather than a small number in a short cycle is a very limited version of job enrichment' (1990: 102). In any case, Womack *et al.* suggest, the Volvo system is 'almost certain to be uncompetitive with lean production'. Overall, the authors of the widely quoted *The Machine that Changed the World* are unequivocal in their judgement about Lean Production. Quite simply they state it is a 'better way of making things which the whole world should adopt . . . as quickly as possible' (1990: 225).

Negative outcomes have, however, also been reported. JIT and LP are seen by the pessimistic 'realists' as leading to job losses, work intensification, higher levels of stress on workers, longer hours (in so far as the models are often built on Japanese practice, critics not unreasonably point out the realities of work in that country), and a bearing-down on all forms of 'slack' in the system which workers might formerly have used to gain a respite. For example, Turnbull (1988: 18) observes, the 'deleterious effects of JIT affect all and sundry'. Garrahan and Stewart (1992) are critical of the intrusiveness of the 'Neighbourhood Check' scheme at Nissan. This is in tune with the tenet of Lean Production which seeks to trace defects quickly to their source.

One of the most well-elaborated critiques of the concept of 'Lean Production' is that mounted by Williams *et al.* (1992). These authors argue that the Japanese producers to whom this label has been attributed do not in fact represent a different 'system' or 'era' which is gradually diffusing across the globe and across different industries. Rather, the series of innovations is seen as simply part of a 'trajectory' (1992: 329). Data produced by Williams *et al.* show that over the long term what has been occurring is a Japanese catching-up of the Americans in automobile production. Toyota's added-value per employee by 1987 was 28% better than Ford's but Nissan's was 18% worse than Ford's. What the Japanese have mainly been doing is taking labour out of each unit of output on a continuous, year-by-year basis.

Another important caveat made by critics is that the comparisons are often flawed because, in the main, the Japanese plants are less vertically integrated than Western ones. The Japanese car producers are largely assemblers who draw heavily on sub-contractors. In contrast, the American plants and companies are much more than assemblers of components, hence plant-to-plant comparisons are misleading. Moreover, when the extreme individual cases of particular plants are set aside, 'the corrected build hours required for weld, paint and final assembly in average Japanese and American plants may not be very different' (Williams *et al.*, 1992: 332).

The controversy surrounding Lean Production seems set to continue. For example, on the one hand, as a perceived alternative to the Fordist form of production it is claimed that LP offers opportunities for front-line workers to recapture many of the competencies and responsibilities formerly ceded to specialists. On the other hand, there are the alleged downside effects of job losses, work intensification and stress (Parker and Slaughter, 1988b). As a recent study by the International Labour Office observes,

> Only if the promising results of lean production can be accomplished without its more ugly face will the concept maintain its attractiveness and mark the big breakthrough which the MIT study (*The Machine that Changed the World*) credits it with; and only then will there be justification of a synthesis of greater efficiency and more fulfilling and humane work of which some of the European unions, and also some managers, dream when they address the concept.
>
> (International Labour Office, 1993: v)

Other criticisms are that higher productivity levels are largely attributable to higher capacity utilization in Japanese plants; that Japanese workers work longer hours and take fewer holidays than Western equivalents. In manufacturing as a whole, average annual working hours for production workers in 1989 were 2,189 in Japan compared with 1,642 in West Germany, that is a difference of 547 hours (33% more than in Germany) (Japanese Ministry of Labour, 1990: 227).

A rare, detailed, empirical study of Lean Production in German industry reaches fairly pessimistic conclusions or, as the author prefers, 'realistic' conclusions (Cooke, 1993: 92). Essentially, he finds, Lean Production is the 'ultimate cost-cutting weapon . . . all the rest is a sophisticated framework within which costs are kept down'. As a result, on balance, workers 'stand to lose more than they gain from the impact of LP in the workplace'. Few will achieve the enrichment promised by the rhetoric of LP. In addition, job losses must occur. There will also be deleterious consequences even for the middle-sized firms which on the surface appear to gain most from the increased outsourcing. Cooke's studies suggest they will be subject to the cost-cutting and quality-raising imperatives of the large-firm customer.

There has been rather less critical comment about TQM. Wilkinson *et al.* (1992: 17) note, however, that TQM does not always enhance employee involvement in the way normally claimed for it. Rather than being enabling it can also increase the degree and incidence of monitoring and control. 'TQM ideas can also be used to reinforce a management style rooted in Taylorism' (Wilkinson *et al.*, 1992: 18).

Sewell and Wilkinson (1991) contend that JIT and TQM *extend* management control. They draw on Foucault's work to suggest that the technology of these two methodologies enables a 'disembodied eye' to 'overcome the constraints of architecture and space to bring its disciplinary gaze to bear at the heart of the labour process' (Sewell and Wilkinson, 1991: 19). Two sets of disciplinary forces are noted: the first is top-down and stems out of increasingly powerful management information systems, the second is horizontal and arises out of scrutiny from co-workers in manufacturing cells. Sewell and Wilkinson suggest that these surveillance attributes are 'deliberately designed' into JIT/TQM.

In judging the outcomes and consequences of the new work methods it is important to remember that the systems which are being superseded are not without problems of their own. Traditional production processes in the electrical and metal product industries, for example, were characterized by job fragmentation and short job cycles. These characteristics resulted in a number of well-recorded physical as well as psychological problems (Seppala, 1989). The new FMS and cellular structures have had 'a strong impact on the control, autonomy, responsibility and knowledge demands at different levels of the organizations' (Seppala, 1989: 300). The advantageous outcomes are seen arising from the fact that a greater number of people in these organizations have the opportunity to have knowledge of, and maintain direct contact with, a wider range of people including customers, suppliers and contractors. Enhanced control and autonomy are seen to be associated with the new group work and the result can be more meaningful work. For the individual it can also offer learning opportunities.

The question of the *employment implications* of the automated 'factory of the future' still provokes controversy. Forester refers to the 'utopian visions of automated factories' which overlook the 'high cost of high tech and the enormous complexity of factory operations' (Forester, 1989: 10). On the other hand, the vastly reduced number of workers to be found in today's modern factories is all too evident.

One of the issues attracting most controversy has been concerned with the implications of new manufacturing methods for 'skills'. What will be the impact on skill levels and on the range and types of skills required? The history of the new technology debate has been witness to two main (conflicting) theses. On the one hand the 'de-skilling' thesis propounded the view that the advance of new technology or at least the way in which new technologies were designed and implemented would tend (and was tending) to erode traditional levels of skill (Braverman, 1974; Nichols and Beynon, 1977; Wilkinson, 1983). The underlying idea that increasing automation undermines skill requirements goes back to Bright (1958). His model showed an inexorable fall in 'level of required worker contribution' as mechanization progressed. But Adler (1986) uses aggregate census data to show that, on a decade-by-decade basis, the changing occupational structure reveals an upgrading effect.

The popularity of the notion that new manufacturing technology de-skills has, in part, been attributed to the sales pitch of vendors. These players like to claim that investment in their machines will be repaid through a reduced need for labour and also through the way in which it supposedly permits the displacement of expensive skilled labour in favour of cheaper unskilled labour. Studies by Buchanan and Boddy (1983) and by Adler (1986) suggest, however, that, on the contrary, higher skill levels are required. Huczynski and Buchanan (1991: 342) comment that 'sophisticated, flexible, expensive equipment needs sophisticated, flexible, expensive people to operate it effectively'. Adler makes the point that the 'myth' of de-skilling presents a 'major obstacle to effective planning for the implementation of new technologies' (1986: 11).

BARRIERS AND PREREQUISITES

A number of factors can be identified as typical *barriers* to the successful implementation of new manufacturing methods. Looked at from the other direction, these same problems can be seen as constituting the list of things which need to be got right, that is the prerequisites for New Wave Manufacturing. We can therefore appropriately view 'barriers' and *'prerequisites'* as forming a single agenda item in this section. That agenda is constituted by five main items: first, the barrier of treating NWM approaches as merely a series of technical fixes; second, the need to achieve integration; third, the need for premium levels of positive commitment and high levels of competence from employees; fourth, the need for alterations to conventional practice which extend beyond the point of production right through the production chain; and fifth, the need for a change in company culture. In this section we attend to each of these issues in turn.

From the myriad problems identified so far by research, one in particular stands out time and time again. This is the tendency for the majority of companies in the West to treat the new methods in a one-sided technical way. The various 'packages', be they JIT or MRPII or MSE, are viewed as 'bolt-ons' or technical fixes. When they eventually fail to fix, and instead proceed to generate their own new set of problems, disillusionment soon follows. The academic researchers into production and operations have increasingly noted this worrying tendency. They have identified it as an 'implementation' problem and have placed it higher up their agenda than heretofore (Voss, 1988; Aggarwal and Aggarwal, 1985; Kinnie and Staughton, 1991). Practitioners too have been alerted to the issue.

In some instances this has merely meant that behavioural aspects are given greater attention in order to avoid 'teething troubles' during the implementation phase. Others have recognized that in reality the problem goes much deeper and that at issue is a fundamental alteration in the social arrangements of the firm.

It has now become evident that Japanese competitiveness does not stem from technological superiority and higher levels of automation. Recent studies in the USA led to this conclusion: 'The source of flexibility in production systems is their organizational characteristics rather than their technological capabilities. Under flexible production, technological tools become flexible not because they are micro-processor based but because they are implemented in a context where they can be flexibly used' (MacDuffie, 1992: 35, cited in Sengenberger, 1993: 6). This point is not disputed by Womack *et al.* (1990) for they too observe that robotization of car manufacturing has proceeded further in Europe than Japan. In other words, competitiveness derives from the *organization of production* rather than simply from increased use of microprocess technology.

Similar points have been made by Kenney and Florida:

> In sharp contrast to the US experience, the implementation of industrial automation in Japan involves the creation of new work environments and the cultivation of workers' intellectual assets as well as their intellectual skills. Japanese corporations have thus far chosen not to implement computer integrated systems, preferring to have workers and managers experiment with FMS. The shift to automated manufacturing in Japan involves organizational innovation and human intervention as an important complement to technological change.
>
> (Kenney and Florida, 1988: 141)

This statement encapsulates a critical message about the prerequisites and barriers issue and reinforces the central point in the introductory section to this chapter: the social organization of production is of at least equal and arguably of greater importance than AMT in the successful adoption and implementation of the NWM methods.

Another aspect of the problem is the pervasive assumption that installation of one of the new packages essentially entails systematically following a series of relatively simple steps. At best, non-technical issues such as communicating with employees are regarded as one of the tasks to be ticked off along the way. The truth is very different. The new methods such as JIT and MRPII require fundamental changes to the pre-existing organizational structure and procedures. Changes are required in roles, priorities, departmental objectives, ways of operating, control, interdepartmental working and so on.

Organizations 'cannot maintain the same structure, habits, same performance evaluation systems and simply add (say) JIT to the existing practices and events in the organizational system and expect it to work' (Safayeni *et al.*, 1991: 36). The problems or barriers are in part organizational (inappropriate structures and procedures for example), in part to do with Human Resource interventions (a need for new skills and competencies for example), and in part 'attitudinal' (a requirement for a new level of commitment for example). Picking up on this last point, Peters (1987: 118) observed that 'most just-in-time experiments have failed to reach their potential, not because of inadequate computerisation, but because of a fundamental failure on the part of participants to understand the new attitudes of trust, cooperation and mutual investment'. The message, in other words, is that organizational and HR issues are integral rather than peripheral. These elements require further examination.

The problem of integration arises in various ways. One of these is that in complex manufacturing organizations it is rarely the case that just one approach or programme is perceived as paramount and pervasive. More typically a range of methodologies are being pursued simultaneously. As

Safayeni *et al.* (1991: 30) observe, the situation is 'analogous to the medical system where a person with multiple problems may find himself at the mercy of medical experts, often with little attention to the patient as a whole'.

A further problem arises from the fact that the new advanced methods such as CIM, JIT and MSE are usually predicated on premium levels of commitment and competence from employees. Such attributes simply cannot be taken for granted. They represent major challenges in their own right. The need for a 'change in attitudes' is noted by virtually all commentators on the various new manufacturing methods. The point is made graphically by one of the cell leaders interviewed by Oliver (1989: 37). When discussing the finer points of people specification items for future cell membership the statement was made: 'you can almost forget the job specification and write "attitude"'. However, bringing about changes in 'attitude' and 'commitment' is recognized as a major challenge in its own right by social scientists. Merely to add them to the new manufacturing shopping list of needed items is therefore potentially to trivialize a major problem.

One reaction might be to seek to evade it by investing in more advanced technology as a way to reduce dependency upon the Human Resource. This would appear to rest on a misapprehension, for, as Huczynski and Buchanan (1991: 342) observe, 'Advanced Manufacturing Technology makes skill and commitment *more* important, not less.'

Another prerequisite is that if full advantage is to be taken of the possibilities of new manufacturing methods then alterations are required, not only at the point of production, but right up the production chain. These include new production lay-outs, cellular methods, Total Quality processes, redirecting marketing to exploit the flexibility of FMS, new arrangements with suppliers, new design and so on (Bessant and Haywood, 1986). Evidently, the span of the managerial agenda if proper attention is to be paid to the organizational and HR dimensions of New Wave Manufacturing, is wide. The vital point, however, is that it seems the main benefits derive, not from the technical characteristics of the new methods alone, but largely from the organizational changes which accompany the new technical possibilities.

The fifth prerequisite is the apparent need for a new organizational culture. The importance of this has been noted in research by Ansari (1986), Helms *et al.* (1990) and Lee and Ebrahimpour (1984). These different sets of authors set out to identify the critical success factors for the implementation of JIT and similar new wave approaches. From a range of factors, including technical as well as Human Resource issues, the paramountcy of appropriate corporate culture has been emphasized by all of them.

Of course it is one thing to note how Japanese culture is congruent with NWM methods but it is quite another to recommend wholesale adoption of similar cultural attributes. Few authors in fact do so. More typically analysts such as Sohal *et al.* (1989) argue the need to *modify* the methods so as to take account of Western 'labour relations, existing management styles and agreements' (1989: 23).

This point really gets to the heart of what now needs to be done next. We have noted in this introductory chapter the meanings and characteristics of the New Wave Manufacturing methods; we have recorded some of the main reasons for adoption and the key outcomes; we have also discussed barriers and prerequisites for successful adoption and implementation. From all of this it has become clear that the best of the Japanese manufacturers have evidently moved on to a new, qualitatively different, plane. Given the patchy record of many Western companies that have experimented with elements of the new system, the priority now must be to search for a more thorough understanding of the essential meanings of the new system and how they can be configured in Western contexts. Analysis of this kind requires the capability to draw upon the respective strengths of the Production and Operations Management perspective and the behavioural/organizational perspective. This sets the agenda for the rest of this book.

CHAPTERS IN THIS VOLUME: RATIONALE AND OVERVIEW

Those then are the key issues and problems. So what do the chapters in this book add to their understanding and resolution? In Chapter 2 Joe Tidd gets to the heart of the matter by seeking to make explicit the links that exist between corporate strategy, technology and work organization. He uses Michael Porter's three types of strategy (cost, quality and innovation) to structure the analysis. As Tidd points out, superficially there is a clear enough link between manufacturing arrangements (technology/organization) and each of Porter's three business strategies. In simplified form this would suggest that JIT 'fits' with a cost leadership strategy; TQM fits with a strategy based on differentiation by quality; and a multifunction team-based organizational form with FMS would seem to 'fit' with an innovation strategy. This clarification of the broad sorts of interconnections is in itself a valuable contribution by this chapter. However, Tidd goes further. He points out that in the 1990s some of the traditional trade-offs such as that between cost and quality are no longer constraining. Hence, leading manufacturers are now changing their manufacturing strategies. They are able to utilize the space which now exists in manufacturing potential to pursue low-cost production *and* differentiation by quality. Moreover, under these new competitive conditions manufacturing managers need to be even more aware that their traditional organizational assumptions and arrangements may have locked them in to the adoption and development of technologies which are sub-optimal. Under these circumstances a clear lesson emerges: 'firms should focus on organizational change consistent with their chosen strategy before investing in Advanced Manufacturing Technologies'.

In Chapter 3 Nick Kinnie and Roy Staughton tackle the problem of *implementing* manufacturing strategy. Conventionally, a discussion of this kind would be located at the end of the book. But the whole point of this chapter is that implementation issues are so critical that planning for implementation needs to occur at the beginning of the whole NWM process and not be left as an afterthought. This is a message fully in tune with the spirit of this book, Kinnie and Staughton in this chapter make an important amendment to Terry Hill's renowned model. The 'implementation' issues – which are in fact the crucial HR matters of education and training, communications and the revision of work practices – are too important to be left to the end.

In Chapter 4 Arthur Francis makes a further valuable contribution to the debate and to the state of knowledge by focusing on the hitherto rather neglected issue of the design–manufacturing interface. This chapter hammers home the point that superiority in New Wave Manufacturing is not *just* about shopfloor issues of actually making the product. Improving the design–manufacturing interface is a key part of the wider agenda. Francis draws on the latest and most practical ways in which Western manufacturing firms could improve their design and manufacturing activities so that they operate more effectively. However, as his argument unfolds it becomes clear that he believes that these practical changes cannot and will not come about incrementally. He traces the relative failure of Western firms in these matters to deep roots. Furthermore, the nature of the required changes is drawn but he suggests that without extra-firm institutional assistance, individual firms will not feel sufficiently impelled to make the necessary step-change.

Chapters 5 to 10 examine, in detail, each of the main approaches which are usually regarded as constituting the core of the new manufacturing. Harry Boer in Chapter 5 tackles Flexible Manufacturing Systems (FMS). He explains the characteristics and benefits of FMS but the main part of his chapter addresses the organizational factors which are necessary for its operation and success. Boer illustrates his points by drawing on his case research in Belgian, Dutch and British companies as well as from the secondary literature on FMS. He observes that 'FMS is a highly regulating technology. Requiring uniform behaviour as regards data input, parts fixturing, change of worn tools and predictable reactions to disturbances and breakdowns, FMS increases the number of rules and procedures to regulate the employees' behaviour'. He also spells out in detail the organizational prerequisites for successful implementation of FMS. Attention deserves to be drawn in particular to Figure 5.10. Running in parallel with the need for more predictability, FMS also requires that shopfloor workers are 'sufficiently trained and skilled to operate the different elements . . . and to resolve all kinds of problems that may occur'. Different sets of organizational arrangements to allow for this are explored by Boer and he suggests that 'the most feasible is a flexible organization based on a semi-autonomous group of well-trained, multiskilled operatives'. Aspects of this are picked up for further debate by David Buchanan in Chapter 10.

Another thoroughbred in the New Wave Manufacturing stable is the idea and methods of Total Quality Management (TQM). Indeed, this is probably the most well-known methodology to the public at large. Patrick Dawson tackles this subject in Chapter 6. He covers the concept, the practice and the HRM issues. The chapter emphasizes the practical reality of TQM and illustrates the points with a case analysis of Pirelli. The result is an excellent guide to the *idea* of TQM balanced by a much-needed analysis of its operation in practice. One of the observations made by a manager at Pirelli in Australia summarizes a great deal about TQM and indeed other NWM methods. He notes the breakthrough in getting people to talk about manufacturing problems. As he says, TQM 'formalizes this process and creates an environment where they *have* to'. But, it is 'not until you change it to *want* to that you start to get results'. As Dawson points out, TQM requires an effective link between technical and social processes – a theme echoed throughout this book.

In Chapter 7 Malcolm Hill explains the possibilities and problems associated with Computer-Integrated Manufacturing (CIM). CIM refers to the potential for a truly integrated manufacturing effort from product conception and design right through to assembly and after-sales service using a common system and common database. The chapter highlights the Human Resource issues critical to the success of CIM, namely, the way labour is deployed and organized, and the importance of training to resolve technical problems, and industrial relations. Especially noteworthy is Hill's point about the need to train managers in the possibilities and practicalities of CIM.

Information drawn from five case companies which were pursuing CIM suggests that the main efforts have been directed towards achieving improvements in the separate areas of design and/or manufacture rather than full integration between these. In the light of the major themes analysed in this book it is important to note Hill's conclusion as to why this progress to full integration has been impeded. He suggests that, 'contrary to some of the literature' (and he cites Forester *et al.*, 1992) the rate of progress on integration is 'shaped more by market pressures and technological constraints than industrial relations, training or organizational issues'. He does go on to observe, however, that 'Human Resources' of the requisite quantity and quality are clearly required to solve these technical and market problems'. A point worth pondering is whether this observation about technological and market factors can be extended beyond CIM to other NWM innovations. The weight of evidence would suggest that this conclusion is rather more specific to CIM *per se*.

In Chapter 8 Alan Spreadbury explains the meaning and operation of Manufacturing Resource Planning (MRPII). As he observes, at a mechanical level MRPII is essentially simply a computerized ordering and tracking system which offers the basis for devising improvements. But, when coupled with appropriate HR strategies, the method becomes a highly powerful tool. Drawing upon detailed insights into the implementation of MRPII in a British affiliate of the American pharmaceutical company Eli Lilly, along with other cases, the chapter gives one of the most rounded accounts of the reality of MRPII and of its potentialities and limitations.

The phenomenon of Just-in-Time (JIT) continues to attract great interest – and controversy. In Chapter 9 Alan Harrison provides a thorough analysis of it and demonstrates how it can be turned into a 'formidable competitive weapon'. The chapter explains clearly the techniques of JIT and methods of implementation. Of particular note is the discussion concerning how 'Japanese methods' of this sort can be 'put to work in a Western setting'. Harrison takes issue with Turnbull that industrial relations (IR) considerations have a serious impact on JIT in Britain but he does, none the less, argue that new methods 'cannot simply be overlaid on to existing management practices' – and in this connection he wonders whether even Rover's New Deal which, unusually, offers employment security, is sufficient to change the HR/IR base. In a valuable final section, Harrison identifies the HR package which he believes is needed to support JIT.

One of the central ideas in the New Wave Manufacturing package is that of re-organization into (relatively) self-contained 'manufacturing cells'. As David Buchanan in Chapter 10 points out, 'Cellular Manufacture' is often misunderstood. Two strands are identified: the *social* arrangement of 'autonomous groups' and the *physical* arrangement of machines in so-called 'Group Technology'. There are clear possibilities for bringing these two different sort of 'groupings' together but they are not always coterminous. The proliferation of 'teamworking' is readily apparent to anyone who regularly visits factories today. This chapter disentangles the different types of teams and their associated characteristics. A point of special note is the contrast which is drawn between the Japanese notion of teamworking and what might be termed the Anglo-Scandinavian version. As Buchanan observes, 'teamwork in Japan is not the same as autonomous group working' as normally understood in the West. This, as Buchanan makes clear, is only the start of the argument and a fascinating research agenda is opened up concerning the comparisons between the autonomous work group concept and the Toyota system.

The strand of this argument is clearly continued in Chapter 11 by Paul Adler. Adler is one of the foremost scholars of New Wave Manufacturing methods in the USA. He has researched extensively on worker reactions and the managerial implications. In this chapter he focuses in particular on workers' reactions to Flexible Manufacturing Systems (thus continuing the discussion on the characteristics of FMS raised in Chapter 5 above). Previous research has shown that FMS workers suffered from lack of autonomy and control. But in new case study research reported here, Adler paints a more complex picture. In two new installations he finds workers reporting high levels of job satisfaction and motivation – despite the fact that one of these was organized along conventional lines and the other used semi-autonomous work groups. This leads Adler to hypothesize that the hitherto central concept of 'autonomy' may be leading researchers astray. The essential factor behind motivation and satisfaction may, he suggests, be the efficacy of the job design configuration for the kinds of tasks to be accomplished. Adler is led to a more contingent view of the relationship between job design, satisfaction and worker orientations.

In the final chapter it is acknowledged that while the chapters to this book have added to the debate about new manufacturing methods, there is still a great deal of work to be done. This concluding chapter points up what we see as the most significant issues and the most promising avenues for future research.

REFERENCES

Adler, P. (1986) New technologies, new skills, *California Management Review*, Vol. 29, no. 1, pp. 9–28.

Aggarwal, S. C. and Aggarwal, S. (1985) The management of manufacturing operations: an appraisal of recent developments, *International Journal of Operations and Production Management*, Vol. 3, no. 5, pp. 21–38.

Ansari, A. (1986) Identifying factors critical to success in implementing Just-in-Time, *Industrial Engineering*, Vol. 18, no. 10, pp. 42–52.

Bessant, J. and Haywood, B. (1986) Flexibility in manufacturing systems, *Omega*, Vol. 14, no. 6, pp. 465–73.

Boddy, D. and Buchanan, D. (1986) *Managing New Technology*, Blackwell, Oxford.

Bratton, J. (1992) *Japanization at Work*, Macmillan, Basingstoke.

Braverman, H. (1974) *Labour and Monopoly Capital: The Degradation of Work in the Twentieth Century*, Monthly Review Press, New York.

Bright, J. (1985) Does automation raise skill requirements? *Harvard Business Review*, July/August, pp. 170–74.

Buchanan, D. and Boddy, D. (1983) *Organizations in the Computer Age: Technological Imperatives and Strategic Choice*, Gower, Aldershot.

Cheng, T. (1988) Just in Time Production: a survey of its developments and perception in Hong Kong industry, *OMEGA*, Vol. 16, no. 1, pp. 25–32.

Clegg, S. (1992) Modernist and postmodernist organizations, in G. Salaman (ed.) *Human Resource Strategies*, Sage, London.

Cooke, P. (1993) The experiences of German engineering firms in applying lean production methods, in International Labour Organization (ed.) *Lean Production and Beyond*, ILO, Geneva.

Drucker, P. E. (1990) The emerging theory of manufacturing, *Harvard Business Review*, May/June, pp. 94–102.

Forester, T. (ed.) (1989) *Computers in the Human Context*, Blackwell, Oxford.

Garrahan, P. and Stewart, P. (1991) Work organizations in transition: the human resource management implications of the 'Nissan Way', *Human Resource Management Journal*, Vol. 2, no. 2, pp. 46–62.

Garrahan, P. and Stewart, P. (1992) *The Nissan Enigma*, Mansell, London.

Gerwin, D. and Kolodny, H. (1992) *The Management of Advanced Manufacturing Technology: Strategy, Organization and Innovation*, John Wiley, New York.

Graham, I. (1988) Japanization is mythology, *Industrial Relations Journal*, Vol. 29, no. 1, pp. 69–75.

Helms, M., Thibadoux, G., Haynes, P. and Pauley, P. (1990) Meeting the human resource challenges of JIT through management development, *Journal of Management Development*, Vol. 9, no. 3, pp. 28–34.

Hill, T. (1986) *Manufacturing Strategy*, Macmillan, Basingstoke.

Hill, T. (1991) *Production/Operations Management: Text and Cases* (2nd edn), Prentice Hall, Hemel Hempstead.

Huczynski A. and Buchanan, D. (1991) *Organizational Behaviour*, Prentice Hall, Hemel Hempstead.
Imai, M. (1986) *Kaizen: The Key to Japan's Competitive Success*, McGraw Hill, New York.
International Labour Office (1993) *Lean Production and Beyond: Labour Aspects of a New Production Concept*, ILO, Geneva.
Japanese Ministry of Labour (1990) *White Paper on Labour*, Ministry of Labour, Tokyo.
Karmarkar, U. (1989) Getting control of Just-in-Time, *Harvard Business Review*, Vol. 67, September/October, pp. 122–31.
Kenney, M. and Florida, R. (1988) Beyond mass production: production and the labour process in Japan, *Politics and Society*, Vol. 16, no. 1, pp. 121–58.
Kenney, M. and Florida, M. (1993) *Beyond Mass Production: The Japanese System and its Transfer to the US*, Oxford University Press, New York.
Kinnie, N. J. and Staughton, R. V. W. (1991) Implementing manufacturing strategy: the human resource management contribution, *International Journal of Operations and Production Management*, Vol. 11, no. 19, pp. 24–40.
Lee, S. M. and Ebrahimpour, M. (1984) Just-in-Time production system: some requirements for implementation, *International Journal of Operations and Production Management*, Vol. 4, no. 4, pp. 3–15.
Lincoln, J., Hanada, M. and McBridge, K. (1986) Organizational structures in Japanese and US manufacturing, *Administrative Science Quarterly*, Vol. 31, pp. 338–64.
Lucas Engineering (1991) *The Lucas Manufacturing Systems Handbook*, Lucas Engineering Systems Ltd, Solihull.
MacDuffie, J. P. (1992) Beyond mass production: organizational flexibility and manufacturing performance in the world auto industry, University of Pennsylvania (Wharton School mimeo), Philadelphia, PA.
Meredith, J. R. (1987) The strategic advantages of the factory of the future, *California Management Review*, Vol. 29, no. 3, pp. 27–41.
Monden, Y. (1983) *The Toyota Production System: A Practical Approach to Production Management*, Industrial Engineering and Management Press, Atlanta.
Nichols, T. and Beynon, H. (1977) *Living with Capitalism: Class Relations and the Modern Factory*, Routledge, London.
Noori, H. (1990) *Managing the Dynamics of New Technology: Issues in Manufacturing Management*, Prentice Hall, Englewood Cliffs, NJ.
Oliver, N. (1989) Human factors in the implementation of Just-in-Time production, *International Journal of Operations Management*, Vol. 10, no. 4, pp. 32–40.
Oliver, N. and Wilkinson, B. (1992) *The Japanization of British Industry* (2nd edn), Blackwell, Oxford.
Parker, M. and Slaughter, J. (1988a) *Choosing Sides: Unions and the Team Concept*, South End Press, Boston.
Parker, M. and Slaughter, J. (1988b) Management by stress, *Technology Review*, October, pp. 37–44.
Peters, T. (1987) *Thriving on Chaos*, Alfred Knopf, New York.
Piore, M. and Sabel, C. (1984) *The Second Industrial Divide*, Basic Books, New York.
Roobeck, A. (1987) The crisis in Fordism and the rise of a new technological paradigm, *Futures*, Vol. 19, no. 2, pp. 129–54.

Safayeni, F. *et al.* (1991) Difficulties of Just-in-Time implementation: a classification scheme, *International Journal of Operations and Production Management*, Vol. 11, no. 7, pp. 27–36.
Schonberger, R. J. (1982) *Japanese Manufacturing Techniques*, Free Press, New York.
Schonberger, R. J. (1986) *World Class Manufacturing*, Free Press, New York.
Sengenberger, W. (1993) Lean production – the way of working and production in the future? in ILO, *Lean Production and Beyond*, ILO, Geneva.
Seppala, P. (1989) Semi-autonomous work groups and worker control, in S. I. Sauter (ed.) *Job Control and Worker Health*, Wiley, Chichester.
Sewell, G. and Wilkinson, B. (1991) Someone to watch over me: surveillance, discipline and the Just-in-Time Labour Process. Mimeo, UMIST, Manchester.
Skinner, W. (1985) *Manufacturing: The Formidable Competitive Weapon*, Wiley, New York.
Sohal, A. S., Keller, A. Z. and Fouad, R. H. (1989) A review of the literature relating to JIT, *International Journal of Operations Management*, Vol. 9, no. 3, pp. 15–25.
Turnbull, P. J. (1986) The 'Japanization' of production and industrial relations at Lucas Electrical, *Industrial Relations Journal*,, Vol. 17, pp. 193–206.
Turnbull, P. J. (1988) The limits to Japanization: Just-in-Time labour relations in the UK automotive industry, *New Technology, Work and Employment*, Vol. 3, pp. 7–20.
Voss, C. (1986) *Implementing Advanced Manufacturing Technology – A Manufacturing Strategy Perspective*, IFS, Bedford.
Voss, C. (1988) Implementation: a key issue in manufacturing technology – the need for a field of study, *Research Policy*, Vol. 17, pp. 55–63.
Voss, C. A. and Robinson, S. J. (1987) Application of Just-in-Time manufacturing techniques in the United Kingdom, *International Journal of Operations and Production Management*, Vol. 7, no. 4, pp. 46–52.
Wall, T. D., Clegg, C. W. and Kemp, N. J. (1987) *The Human Side of Manufacturing Technology*, Wiley, New York.
Warner, M., Wobbe, W. and Brodner, P. (eds.) (1990) *New Technology and Manufacturing Management: Strategic Choices for Flexible Production Systems*, Wiley, Chichester.
Westbrook, R. (1987) Time to forget Just-in-Time? Observations on a visit to Japan, *International Journal of Operations Management*, Vol. 8, no. 4, pp. 5–21.
Wilkinson, A. *et al.* (1992) Total quality management and employee involvement, *Human Resource Management Journal*, Vol. 2, no. 4, pp. 1–20.
Wilkinson, B. (1983) *The Shopfloor Politics of New Technology*, Heinemann, London.
Williams, K., Haslam, C., Williams, J. and Cutler, T. (1992) Against lean production, *Economy and Society*, Vol. 21, no. 3, pp. 321–54.
Womack, J. P., Jones, D. J. and Roos, D. (1990) *The Machine that Changed the World*, Rawson Associates, New York.

2

THE LINK BETWEEN MANUFACTURING STRATEGY, ORGANIZATION AND TECHNOLOGY

Joe Tidd

Manufacturing Strategy is dynamic. Firms will pursue different strategies at different times. During the 1970s American and European manufacturers faced intense competition from low-wage Asian companies. Exports from South Korea, Taiwan and Hong Kong grew by more than 10% each year. Therefore Western manufacturers concentrated on increasing productivity. In the 1980s the growing affluence of consumers in the developed economies and the example set by Japanese manufacturers forced many firms to focus on improving the quality of products. In the 1990s the basis of competition appears to have changed once again, and manufacturers in the developed economies are increasingly concerned with reducing the lead time of product development and offering a wider range of products. Each change of strategy has been accompanied by changes in manufacturing technology and organization.

European, American and Japanese manufacturers all made significant investments in new manufacturing technologies, specifically robotics and Flexible Manufacturing Systems (FMS). However, in many cases these technologies have failed to deliver the predicted benefits. Consequently more recent efforts have involved organizational innovations, such as Total Quality Management (TQM) and Just-in-Time (JIT) scheduling. This chapter examines the relationship between strategy, technology and organization. The conventional view is that strategy, technology and organization should be consistent. However, we will argue that recent developments in manufacturing technology and work organization may have overcome many of the traditional trade-offs of manufacturing, thereby allowing firms more 'strategic degrees of freedom'. The chapter begins with a framework which links Manufacturing Strategy, technology and organization, and then examines the technological and organizational implications of strategies based on cost leadership, differentiation by quality and, finally, differentiation by innovation.

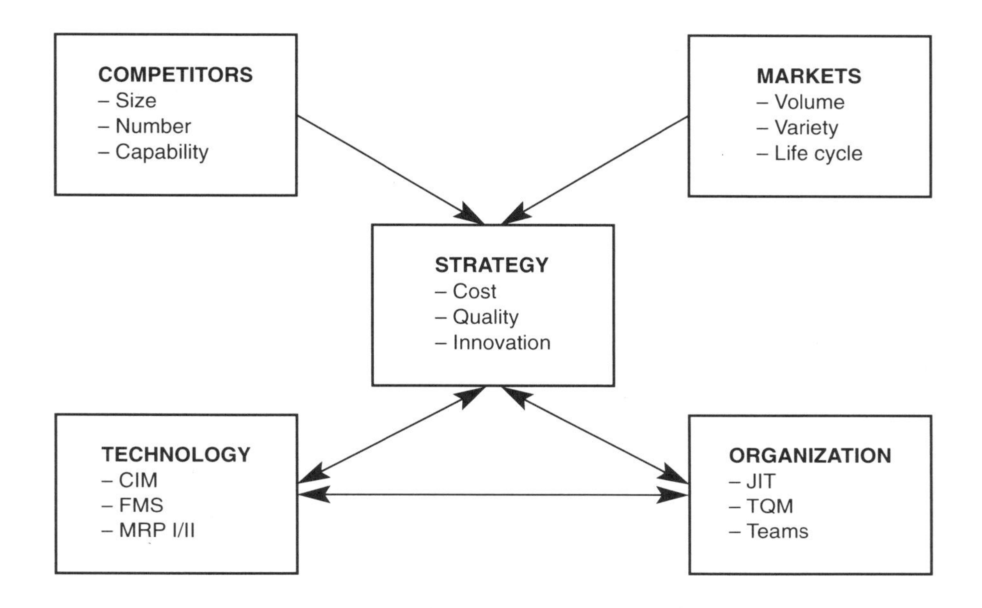

Figure 2.1 An integrative framework for linking strategy, technology and organization

AN INTEGRATIVE FRAMEWORK

The traditional rational model of strategy formulation is based on a 'top-down' approach: corporate strategy is derived from an analysis of market opportunities and competitive threats, and the strengths and weaknesses of the firm. The chosen strategy then determines the appropriate manufacturing technology and organization. In contrast the behavioural model of strategy formulation is based on a 'bottom-up' approach: the goals of different groups within the firm, for example the marketing or finance functions, will influence and constrain the choice of strategy and technology. Both approaches suggest that Manufacturing Strategy, technology and organization should be consistent (Figure 2.1).

Porter (1980) identifies two generic strategies: cost leadership and differentiation. A firm pursuing a pure cost leadership strategy aims to gain a competitive advantage by becoming the lowest-cost producer in an industry. Thus the emphasis is on efficiency, high productivity and economies of scale. This suggests special-purpose automation, long production runs and a limited product range. In contrast, a firm pursuing a differentiation strategy aims to gain a competitive advantage by offering a unique product or service. Differentiation can be on the basis of design, quality, reliability, performance or any other non-price factor. This suggests more flexible forms of automation, small-batch production and a wide product range. We will focus on differentiation by quality, and differentiation by innovation. The choice of strategy will depend on which are the most important order-winning criteria in a particular case. The distinction between order-qualifying and order-winning criteria is important. The former are demanded as basic requirements which must be satisfied, the latter are the source of competitive advantage. For example, in the aerospace industry all aircraft are expected to be able to fly safely, but some customers will demand fuel economy while others may favour low maintenance. In most companies corporate strategy is determined by marketing requirements to which manufacturing must then respond. However, investments in manufacturing resources are long term, and therefore changes in the demands of marketing can result in a mismatch between the two functions. Therefore the manufacturing function should be involved in the formulation of corporate strategy (Hill, 1985). Manufacturing must be able to support a product over its entire life cycle, and therefore Manufacturing Strategy should begin with a review of current and future products.

Production volume, product variety and the dominant form of competition will vary over the product life cycle. Typically, initial production volumes will be low and a wide range of competing designs may exist. Competition will be based on differences in product design and performance. However, as soon as a dominant design has become established, competition will be based increasingly on product quality and reliability. Production volumes will increase and product, production technology and work organization will become more standardized. Finally, the product matures and the production process becomes highly standardized, and competition is based on differences in price. Each stage in the product life cycle will require different manufacturing technologies and organization. Typically, production technology and work organization will need to be most flexible at the early stages of the product life cycle, before a dominant design has emerged, but will become more specialized as the product matures and becomes more standardized (Abernathy, 1978). Many products follow this general pattern. For example, steam, electric and combustion-powered cars competed for customers in the late nineteenth century, and were built to customer specification by craftsmen. The dominant design emerged in the early 1890s, and consisted of a combustion engine in the front driving the rear wheels. Consequently reliability and quality became more important to buyers. Some twenty years later, Ford standardized the product design and production process and became the first automobile manufacturer to compete on the basis of cost.

Different sectors are characterized by different degrees of product and process standardization (Hayes and Wheelwright, 1984). For example, capital goods manufacture is characterized by relatively low-volume, high-variety production, and demands versatile, high-performance manufacturing technologies; on the other hand, automobile manufacture requires high-volume, low-cost production and therefore efficient but relatively inflexible technologies. Normally a firm will attempt to match its choice of production technology and work organization with existing market demands in this way, but alternatively may consciously choose to compete on a different basis in order to gain a competitive advantage.

For example, machine tools are normally manufactured in small to medium-sized batches using general-purpose equipment, but in the late 1970s Japanese manufacturers of Numerically Controlled machine tools adopted a successful strategy based on low-cost, standardized machines. Conversely, the volume car market has traditionally competed on the basis of price, and therefore manufacturers have concentrated on cost reduction through mass-production. However, during the 1980s, Japanese automobile manufacturers increased the flexibility of production in order to be able to offer a wide range of products with short life cycles. Clearly Manufacturing Strategy is influenced by industry norms, but not fully determined by the stage in the product life cycle. Nevertheless the choice of manufacturing technology should be consistent with corporate strategy.

Similarly, technology and organization should be consistent. Early research indicated that technology was a major determinant of organizational structure (Woodward, 1965; Child and Mansfield, 1972), but more recent studies of the adoption of Advanced Manufacturing Technologies suggest that a wide range of options may exist. For example, robotic assembly cells may lend themselves to the use of work groups and other 'Japanese'-style working practices. But it is equally possible that the necessary control, operating and support tasks will be organized into isolated and trivial jobs (Fleck, 1987). Many British firms have adopted sophisticated machine vision and diagnostic systems rather than train workers in Design for Manufacture and quality management. In contrast, Japanese manufacturers have been able to adopt less complex manufacturing technology because they have simplified the organization of production prior to automation (Tidd, 1990). It appears that Advanced Manufacturing Technologies allow a larger 'design space' for matching technology and organization (Bessant, 1991) and a wide range of 'strategic options' (Buchanan and Boddy, 1983). We will consider the technological and organizational implications of strategies of cost leadership, differentiation by quality and differentiation based on innovation.

COST LEADERSHIP

Economic theory predicts that the 'optimum firm' will operate at the scale at which organization and technology produce the lowest average cost. In practice, large firms achieve managerial and financial economies of scale, and benefit from lower purchasing, distribution and sales costs. However, technical economies of scale are usually more important at the factory or plant level. For example, in process-based sectors such as the chemical industry there is a rule of thumb that each doubling of capacity leads to a six times fall in costs. Historically, technical economies of scale have been the result of the specialization of manufacturing technology and organization.

The production of the Ford Model T was the archetype of high-volume, standardized manufacturing. Through a combination of a standard design, extreme division of labour and special-purpose automation, Ford reduced the time to assemble a car by three-quarters, and the cost by two-thirds. Ford was the first to adopt a strategy based on cost leadership, and aimed to produce the lowest-priced car and to use continuing price reductions to produce even greater demand. In 1923 annual production of the Model T reached a peak of almost two million, but just two years later annual sales fell to 800,000. Ford's share of the American market fell from 55% to only 15%, forcing Ford to replace the Model T. This required a complete plant shutdown, the scrapping of a quarter of the existing 32,000 machine tools, and the purchase of 45,000 new tools for the new model. According to one historian,

> mass production as Ford had made it and defined it was, for all intent and purposes, dead by 1926. Ford and his production experts had driven mass production into a deep cul-de-sac. . . . Automotive consumption in the late 1920s called for a new kind of mass production, a system that could accommodate change and was no longer wedded to the idea of maximum production at minimum cost.
>
> (Hounshell, 1984: pp. 12–13)

Nevertheless, mass-production has continued to dominate the automobile industry and many other sectors. A survey of American manufacturers suggests mass-production accounts for almost three-quarters of the output of the automobile industry, and around a third of the output of the electrical and metal products sectors (Ayres and Miller, 1985). However the saturation of mass markets, intense international competition and changing consumer tastes threatened mass-production during the 1970s, resulting in a growing mismatch between the marketing requirements and manufacturing capabilities. Many manufacturers responded by increasing their product range and the number of options available. This satisfied market demands but increased the cost and complexity of manufacturing (Hill, 1980). Consequently in the 1980s many American and European manufacturers reduced their product ranges in order to simplify production and facilitate automation: 'companies, including manufacturers of appliances, autos, copiers, and cameras are following this approach of simplifying and focusing product offerings' (Schonberger, 1987: p. 97).

For example, in the automobile industry Fiat reduced the number of chassis it manufactured from nine to five, and the number of body styles from nine to six. Overall, the number of different models produced by all European car manufacturers fell from forty-nine to forty-three between 1982 and 1990. Despite this, European and American manufacturers failed to match the productivity of their Japanese competitors. The best Japanese plants are twice as productive as the worst North American and European automakers (Womack, Jones and Roos, 1990). Around a third of the productivity difference between the best and worst performing plants is attributable to Design for Manufacture, another third to the level of automation, and the remainder to 'Lean Production', which includes teamworking, Total Quality Management (TQM), and Just-in-Time (JIT) scheduling. This clearly demonstrates the importance of manufacturing technology and organization.

Technology

It is unclear whether Advanced Manufacturing Technology will have the greatest impact in batch or volume production. Some argue that programmable manufacturing technologies will allow smaller firms to gain some of the benefits of automation and therefore challenge larger firms. Others predict that 'economies of scope' will replace 'economies of scale' in larger firms, allowing them to compete with smaller-batch manufacturers. However, the evidence to date suggests that the greatest impact has been in high-volume, high-variety production.

Conventional automation is only suited to high-volume, standardized production, and typically is limited to a few product styles manufactured in annual production volumes of a million. Consequently around half of all products are manufactured on assembly lines, but only 5% are fully automated. Advanced Manufacturing technology appears to have had a limited impact on productivity in those sectors characterized by mass-production. In the automobile industry there is only a weak correlation between automation and productivity. Japanese plants do appear to have consistently higher levels of automation and productivity, but the relationship is less clear from the performance of European and American manufacturers. For example, in Figure 2.2, there is a small cluster of plants which have automated around 40% of production, but have different productivity: the European plant takes more than forty hours to build a vehicle; the American plant about twenty-five hours; and the Japanese plant takes less than twenty hours. This suggests that the productivity advantage of Japanese manufacturers is not simply due to higher investment in manufacturing technology.

Historically productivity in Japan grew by an average of 5% each year during the 1970s, but this fell to 3% following the widespread adoption of

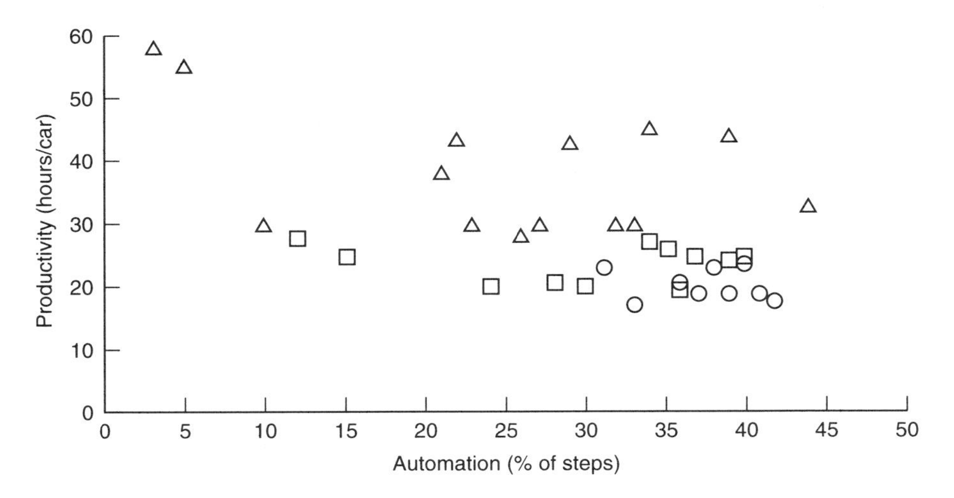

Figure 2.2 Automation and productivity in the automobile industry.
□ = USA; ○ = Japan; Δ = Europe. (From Womack *et al.*, 1990)

Advanced Manufacturing Technologies (Ishitani and Kaya, 1989). In fact the productivity of the Japanese automobile industry peaked in 1980, which is generally considered to be the first year of 'robotization' in Japan (Watanabe, 1987). Between 1980 and 1984 Nissan increased its use of robots almost threefold, but productivity fell by more than 10% over the same period; at Toyota robot use increased fourfold, and productivity fell by just under 5% (Tidd, 1991). This slight fall in productivity is consistent with a change in strategy from cost leadership to differentiation, and appears to be the result of changes in product mix and an increase in the range of products manufactured.

Japanese manufacturers have consistently entered new markets by a strategy of cost leadership by exploiting the experience curve. Empirical evidence from a wide range of sectors confirms that unit costs fall by a predictable amount as the cumulative production volume increases. This is more than simple economies of scale, and is based on continuous improvements in product design, production technology and work organization. In this way manufacturers can anticipate future cost reductions when pricing their products, and therefore increase their market share. For example, there was a strong learning-curve effect in the production of video cassette recorders (VCRs). Between 1975 and 1985 the annual production volume of Japanese manufacturers grew to 35 million units, and unit costs fell by 70%, from 200,000 yen to 60,000 yen. This reduction in cost was achieved by improvements in design and manufacturing: in 1981 a typical VHS-format VCR consisted of 460 mechanical, electrical and electronic components, assembled mainly by hand; in 1983 it had 370 parts, assembled by 52 pick-and-place machines, 11 robots and 9 workers; and by 1985 the standard product had just 250 parts, assembled by 71 pick-and-place machine and 24 robots – representing 98% of all operations (Tidd, 1991). More recently, Korean manufacturers have followed a similar cost leadership strategy, based on their lower cost of labour. However, these products do not directly compete with Japanese VCRs as Japanese manufacturers have increasingly added extra features and increased prices: in 1985 the median price of a Korean VCR was 30% below that of a Japanese machine.

This demonstrates the relationship between strategy and production technology. The relationship between production technology and work organization is equally important. Research in the automobile industry has shown that the level of automation has a significant impact on productivity when plants move from low levels to average levels of manufacturing technology, but that plants with conventional forms of work organization gain much less when they move from average to high levels of automation (Womack *et al.*, 1990). In contrast, plants which use multiskilled teams and 'Japanese' production systems are able to achieve greater increases in productivity by adopting high levels of automation. For example Toyota, which developed Just-in-Time production, has 'ten commandments' of automation (Tidd, 1991):

1. Automation should begin at the end of production, not with the easiest operations.
2. Cost of machines should be in proportion to the value-added.
3. Machines should be easy to start and stop.
4. Machines should be easy to back up in event of breakdown.
5. Machines should stand alone, be easy to relocate, and have variable speed.
6. Machines should be modular and programmable to cope with model changes.
7. New and unproven technology should be avoided; the production line is not a laboratory.
8. Defects should be detected and corrected at source, not passed on to the next process.
9. Production should be pulled from the previous process, not pushed by forced conveyance.
10. Common operations from different lines should not be combined, to ensure maximum flexibility.

These guidelines for the adoption and development of Advanced Manufacturing Technologies illustrate the strong links between strategy and technology on the one hand, and between technology and work organization on the other. Japanese manufacturers place consistent emphasis on cost, quality and flexibility.

Organization

Capacity planning, production control and inventory control are three of the most difficult areas of production management. Capacity planning is a long-term problem, whereas production and inventory control are shorter-term issues. As mass-production is capital intensive, accurate forecasting of demand is crucial. However, the production process is relatively well defined, and consequently production control is almost routine. In contrast, in batch production accurate forecasting is impossible, but production control is critical. Therefore, in mass-production, materials control is triggered from signals generated by stock levels – stock-point generation – and in batch production materials are controlled by signals from sales – order-point generation. In theory, the order-point system is considered to be the ideal to aim for, but the sheer volume of data that this generates means that in practice some form of stock-point control system is most common.

Materials Requirement Planning (MRP) was developed in the 1960s to deal with the problem of inventory control. As the name suggests, MRP systems produce a breakdown of the materials and components required for a specific demand for products. Typically, systems do not work in 'real time', and material requirements are based on forecasts of demand. Consequently, in practice MRP works best for firms with a limited range of products and relatively stable patterns of demand. Manufacturing

Resource Planning (MRPII) was designed to take account of resource constraints and predict the implications of orders on future capacity requirements. These systems may include scheduling, inventory control, activity planning and plant monitoring and control, and aim to reduce cycle time and inventory levels (Meredith, 1987). For more details see Chapter 8 of this volume. However, MRPII systems have been criticized for their complexity and for generating schedules that do not reflect reality on the shopfloor. Consequently many firms have turned to other ways of managing production.

MRP systems attempt to improve inventory control by trying to cope with the complexity of production. Just-in-Time (JIT) scheduling attempts to improve inventory control by reducing the complexity of material and information flows. Therefore, strictly speaking, the two approaches are not incompatible, and hybrid systems do exist. JIT production reduces costs and exposes any inadequacies in the system by eliminating buffer-stocks. The traditional function of buffer-stocks is to reduce uncertainty. Stocks of materials and components reduce the risk of delivery delays and also may allow volume discounts. Work in Progress decouples different stages of production so that isolated disturbances such as machine breakdown do not affect the entire system. Stocks of finished goods act as a buffer against changes in demand. In JIT scheduling material and information flows are matched through the exchange of *kanban*, and materials and components are 'pulled' through the production system by orders, rather than 'pushed' from stock. But JIT scheduling is only one part of a system. It demands high-quality components, reliable machines, rapid machine changeovers, and frequent and reliable delivery of components and materials.

DIFFERENTIATION BY QUALITY

Quality can improve profitability in two ways. Firstly, it can reduce costs by helping to eliminate defects, scrap and rework and by lowering inspection and warranty costs. Typically these may represent up to a quarter of the cost of manufacturing, so the impact on productivity can be significant. This is consistent with a strategy of cost leadership or differentiation by quality. Secondly, difference in relative product quality can increase sales and therefore market share. Research confirms a strong relationship between relative product quality and market share. Moreover, firms with higher quality relative to their competitors are able to demand higher prices (Luchs, 1990). Therefore quality allows a firm to both reduce costs and increase prices, thus improving profitability (Figure 2.3). This is the basis of a strategy based on quality leadership. A large number of firms adopted quality programmes during the 1980s, including major international companies such as Ford, General Motors, IBM, Hewlett Packard, Kodak and Xerox (Shetty and Buehler, 1991). This has involved the adoption of new technologies and techniques, and changes in organization.

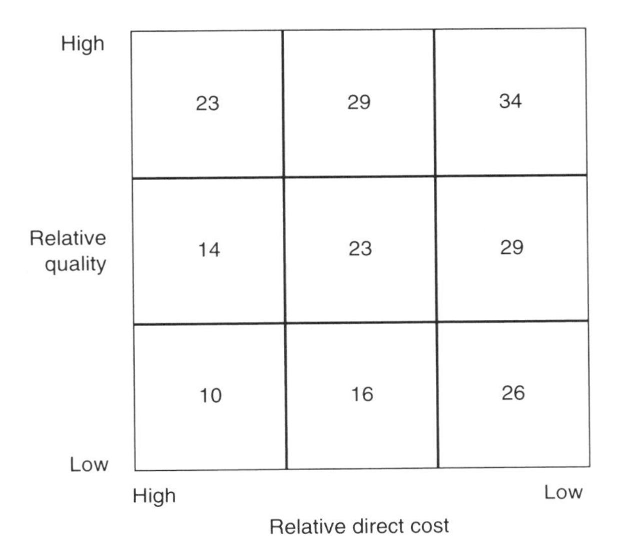

Figure 2.3 Relative product cost, quality and profitability
Numbers = % return on investment. (From Luchs, 1990)

Technology

Quality management is based on techniques rather than technologies. The seven tools of quality control – cause-and-effect diagram, histogram, check-sheet, Pareto diagram, control chart, scatter diagram and graphs – together with the seven 'new' tools – relations diagram, affinity diagram, systematic diagram, matrix diagram, matrix data analysis, process decision program chart and arrow diagram – are the foundations of quality management (Dale and Plunkett, 1990).

Nevertheless technology has begun to make a contribution to quality management, both indirectly and directly. Technology can help to improve quality indirectly by better control of the production process, and directly by automated inspection and testing. In addition it can be used to support the use of the tools of quality control. Software can easily perform the appropriate computations and display the results graphically. However, fully automated sensing and measuring are increasingly used to provide fully automated quality control. The most common application is in-process Statistical Process Control (SPC).

However, the differences in the performance of Japanese manufacturers and their competitors in the West is not the result of more advanced technology. In the USA General Motors invested some $40 billion in its new Saturn plant which features the latest Computer-Integrated Manufacturing technology, but this has failed to achieve the levels of productivity and quality of the NUMMI plant which is a joint venture with Toyota. European and American firms place considerable emphasis on the use of computer-aided statistical process control, whereas the Japanese rely much more on training operators and members of quality circles in the use of a range of tools for quality control (Dale and Tidd, 1991).

Organization

Quality circles are the most obvious example of organization for quality. For example, there are estimated to be some 750,000 quality circles in Japan with 5.5 million members. Consequently many companies in the West adopted quality circles – though with limited success. Too often quality circles are implemented in isolation, and are not fully integrated with the existing organization. In Japan circles are considered to be an integral part of normal working practice, rather than a separate activity. But quality circles are only a small part of quality management.

Most firms have focused on improving the quality of the internal process, and less on external perceptions of quality. Conventional quality circles and improvement teams tend to be based on the shopfloor and are concerned with improving the production process, rather than product design or performance. There is a need to link the organization for quality with a strategy based on quality. More recent developments in quality management have begun to address this problem, and are central to a strategy based on differentiation by quality. For example, Quality Function Deployment (QFD) and Taguchi methods aim to identify the relative importance of different product attributes and focus development on these areas. These demand inputs from a wide range of functions, including engineering, marketing and manufacturing. Such multifunctional teams are essential for a strategy based on quality.

However, by definition, a strategy of differentiation requires customers to perceive some difference among competitors. But in many sectors quality has become an order-qualifying criterion, rather than an order-winning criterion. For example, in the automobile industry a growing number of American and European manufacturers have achieved quality levels similar to their Japanese competitors. But while manufacturers in the West have been preoccupied with matching the Japanese in terms of quality and productivity, Japanese manufacturers have concentrated on developing the capability to offer a high variety of products with shorter life cycles without significantly increasing costs, (Ferdows *et al.*, 1986). Therefore in many cases quality is no longer a sufficient basis for differentiation, and increasingly firms are having to differentiate their products in terms of other non-price criteria.

DIFFERENTIATION BY INNOVATION

One of the most fundamental trade-offs in manufacturing is between efficiency and flexibility, the so-called 'productivity dilemma': product innovation becomes more difficult and less frequent as the efficiency of the production process improves (Abernathy, 1978). Flexible forms of manufacturing technology and organization promise to help overcome the 'productivity dilemma', allowing the economical production of variety, frequent design changeovers and rapid response to design and market changes. This suggests that Advanced Manufacturing Technology will be most appropriate for firms which compete on the basis of differentiation rather than cost or quality. Blois (1985) identifies markets which demand a wide range of products and those which are volatile and unpredictable as being most suited to the application of Advanced Manufacturing Technologies. Similarly Goldhar and Jelinek (1983) argue that the current market environment demands speed and product variety, and therefore firms should increase the speed of market response by using Computer-Integrated Manufacturing technologies, and increase product variety through Flexible Manufacturing Systems.

Technology

There is no generally accepted definition of a Flexible Manufacturing System (FMS), and consequently the technology has become 'all things to all men'. Many predicted that it would revolutionize small-batch production: 'the final major development, in embryonic form in the early 1980s, was the flexible manufacturing system . . . by the 1990s we will surely see the fairly widespread emergence of fully automated production in many sectors including those now characterised by small batch production' (Kaplinsky, 1984). But in practice most FMS are machining systems rather than manufacturing systems. Moreover, programmability and flexibility are not synonymous. A study of FMS in the USA concluded that in most cases systems were relatively inflexible: part numbers were restricted to an average of eight; of all the components manufactured in the plant within a given size envelope only 3–4% could be produced on the FMS; and in the event of the product failing a high proportion of the FMS would become obsolete (Ingersoll Engineers, 1984).

However, significant international variation exists. A comparative study of FMS in the USA and Japan found that systems in Japan produce almost ten times as many variants as those in the USA, 93 compared with 10. In addition, for every new part produced by an FMS in the USA, 22 were introduced in Japan (Jaikumar, 1986). Other surveys of FMS performance reveal similar variations in flexibility: only 17% of FMS in Europe process more than 50 different parts, compared to 37% of systems in the USA, and 49% in Japan (ECE, 1986). Similar differences in performance are found in robotic systems. Japanese users manufacture a wider range of products, with shorter life cycles, than their British counterparts: only a fifth of all products assembled by robot in the UK have life cycles of less than three years, compared with a third of products assembled by robot in Japan (Tidd, 1991). These differences in flexibility appear to be the result of the choice of technology, and Japanese manufacturers favour less complex manufacturing technologies than their counterparts in the West.

FMS flexibility is constrained by the choice of lay-out, hardware and software. The two basic system configurations are the cell and line, although in practice a combination of the two is used. The main factors affecting the choice of configuration will be the scale, complexity and flexibility of production. Generally a line configuration will require less sophisticated technology than the cell because each robotic or machining station will perform a smaller number of tasks and therefore will not need the same degree of dexterity or 'intelligence'. In addition the cost of parts-feeding and fixtures is typically less. A 'high' production volume will suggest a line lay-out, whereas 'low' volume applications suggest a cell. If the production task is complex and cannot easily be broken down into simpler tasks, a sophisticated robotic or machining cell may be necessary rather than a number of simpler machines in line. However, neither the line nor cell configuration is inherently more flexible. Hardware choices include the type of machine tools, handling systems and storage arrangements. In addition the software is often complex and therefore not easily changed. Consequently it is necessary to build in the required flexibility at the design stage:

> The limited flexibility of a flexible machining system can, of course, be a disadvantage unless the system and the parts that it is to make have been carefully chosen. Either the system must be designed for a limited product range for which future demand is predictable or it must have, or be capable of having added, more flexibility at the expense of simplicity and cost. In general, a company installing a flexible machining system must be prepared to reorganise its manufacturing procedures if the full potential of the system is to be realised . . . planning and understanding are crucial because in one way the system is inflexible, as most of the options should be planned at the beginning.
>
> (NEDC, 1984: 20)

Different strategies will demand different kinds of flexibility. There are many types of flexibility, but the three most important are: *mix flexibility*, the capability to produce a wide range of products in small batches; *product flexibility*, the capability to change over between one high-volume product to another efficiently; and *volume flexibility*, the ability to operate efficiently at different production volumes. Mix flexibility may be necessary where competitors offer a full range of products or demand is highly fragmented. Product flexibility is needed in mass markets characterized by short product life cycles, such as consumer electronics and fast-moving consumer goods. Volume flexibility will be desirable where demand is uncertain or cyclical, such as the automobile industry. Each will require different manufacturing technologies and organization.

Organization

A number of studies suggest that around half of the benefits claimed for FMS are in fact the result of organizational and managerial changes (Bessant, 1991). Increased market uncertainty has led many firms to improve the responsiveness of their workforce to changes in demand – so-called numerical flexibility. Also, changes in manufacturing technology often require a different range of skills – so-called functional flexibility. This suggests a model of the 'flexible firm' consisting of a core workforce which is functionally flexible, and peripheral groups of workers to provide numerical flexibility (Atkinson, 1984). The chosen strategy should determine the appropriate balance between core and peripheral groups, and emphasis on functional and numerical labour flexibility, but in practice many firms have acted opportunistically in response to recession. For example, British firms appear to have made greater use of numerical flexibility than functional flexibility. In the UK temporary workers now account for more than 5% of the workforce, similar to Japan (Tidd, 1991). But there is less evidence of any significant increase in functional flexibility.

Traditionally factories are organized on either a functional (process) or line basis, depending on the scale and flexibility of production. In a functional or process lay-out, tools and equipment required for a particular process, for example grinding, are grouped together. This arrangement simplifies control, but increases scheduling and handling problems. Consequently it is most suitable for large-batch production. However, a line layout is more efficient for the mass-production of standardized products. Tools and equipment are arranged in the order of the different stages of production. But for smaller-batch, higher-variety production a group or cell lay-out may be more appropriate. In most plants a small proportion of products and components account for the majority of the resources required, and therefore these may be grouped together and given dedicated resources. By identifying related families of products and components the appropriate tools and equipment can be grouped together. This improves efficiency and machine utilization by reducing set-up and transportation time, but demands significant and organizational change.

Effective group or cellular manufacture demands greater functional flexibility. The functional lay-out encourages specialized functional skills or crafts. The efficiency of line production is based on the extreme division of labour, and therefore demands little skill or training. Group Technology is organized around a narrow set of products or components, but a wide range of equipment. Typically workers will be required to operate a range of smaller machines, rather than be dedicated to one large machine. Moreover, technological developments challenge the historical boundaries between trades or crafts, for example a robotic cell will require electronic, mechanical and hydraulic maintenance. Consequently functional flexibility is most appropriate for high-technology, capital-intensive sectors, whereas numerical flexibility is more commonly needed in low-technology, labour-intensive industries.

In North America and Europe these changes in work organization have often been part of a move towards Japanese-style work practices, and have included multiskilling, group work and quality management. For example, in the UK, an agreement between Nissan and the trade union specifies that

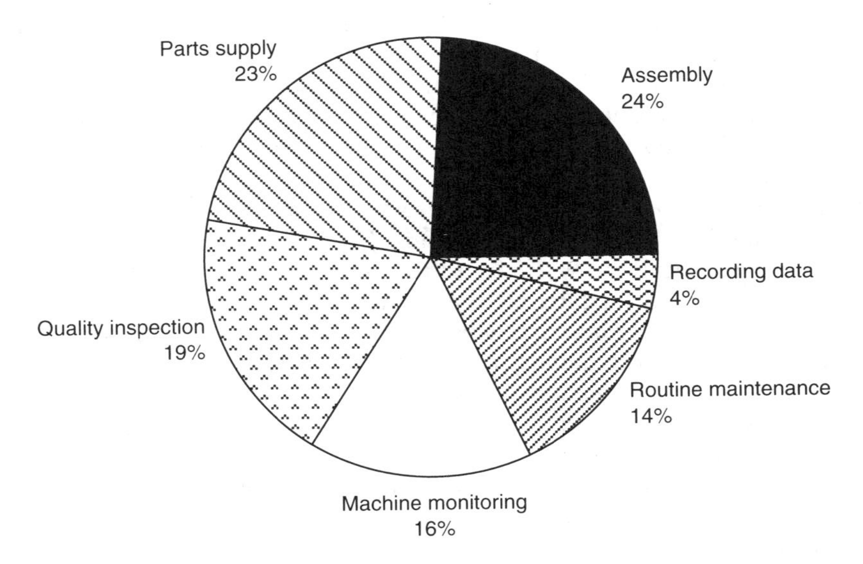

Figure 2.4 Labour flexibility in Japan: tasks performed by machine operators. (From Tidd, 1991)

'to ensure the fullest use of facilities and manpower, there will be complete flexibility and mobility of employees . . . [and] to ensure such flexibility and change, employees will undertake training for all work required by the Company' (Wickens, 1987). Clearly, functional flexibility demands a commitment to training. In Japan it is common for a worker to operate a number of different machines at the same time, and to be responsible for quality control and routine maintenance (Figure 2.4). Similarly, a few British firms have introduced new positions such as 'operator-setter' and 'operator-technician' in an effort to break down the traditional demarcation lines and improve labour flexibility. However, many manufacturers in Europe and North America still have organizations which are unable to exploit fully the strategic potential of Advanced Manufacturing Technologies (Hayes and Wheelwright, 1988).

STRATEGIES FOR THE 1990s

The conventional product and process life cycle model suggests that different strategies will demand different work organization and manufacturing technology. Therefore strategies of cost leadership, quality and innovation have traditionally been incompatible. Indeed recent developments in work organization and manufacturing technology are associated with different strategies (Figure 2.5). However, they also challenge many of the fundamental trade-offs in manufacturing. Cost and quality are no longer conflicting goals, but can be achieved simultaneously to improve profitability. Similarly the historical trade-off between efficiency and flexibility, the so-called 'productivity dilemma', may no longer constrain strategy.

	Cost	Quality	Innovation
CIM	* *	*	* * *
FMS	* *	*	* *
MRP	* *	*	*
TQM	* *	* * *	*
JIT	* * *	* *	*
GT	* *	* *	*

Figure 2.5 Links between strategy, technology and organization
Note: strength of relationship: * = weak; ** = moderate; *** = strong

There is evidence that manufacturers are changing their manufacturing strategies in the 1990s (Figure 2.6). American manufacturers made significant improvements in quality during the 1980s, and are now moving away from their traditional position of high-volume, cost-based competition. European manufacturers have traditionally competed on the basis of design and innovation, but in many sectors have not been competitive in terms of price or quality. European manufacturers began to adopt quality management in the mid-1980s, and are now beginning to compete on the basis of quality. However, this appears to be have been at the price of a loss of flexibility (Tidd, 1991). Most significantly, while manufacturers in the West have been preoccupied with improving productivity and quality, Japanese manufacturers have been building on their past achievements and are developing flexible, low-cost production. This strategy will demand significant technological and organizational innovation.

At first sight different organizational innovations appear to support different strategies: Just-in-Time production is consistent with a strategy of cost leadership; Total Quality Management is consistent with a strategy of differentiation by quality; and flexible, multifunctional teams support a strategy of differentiation by innovation. But the distinctions are not that clear. For example, quality management can support a strategy of cost leadership or differentiation by quality. Similarly, flexible, multiskilled teams are used in both quality management and New Product Development. This suggests considerable scope for matching strategy and organization. It appears that strategy constrains, but does not fully determine organization.

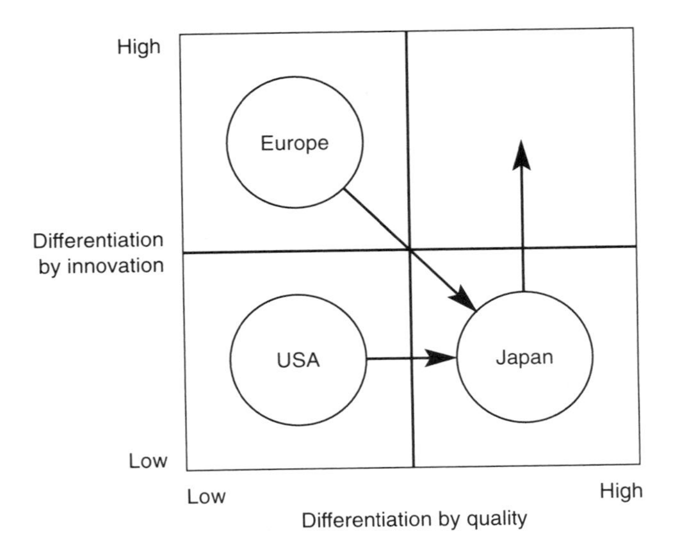

Figure 2.6 Strategies for the 1990s

Similarly, organizational context appears to influence the development and adoption of the technology. For example, in the UK operators often receive little training; communication between design, manufacturing and sales is poor; and relations with suppliers weak. As a result, machines are made 'idiot proof', products are not designed for ease of manufacture, and components are of uncertain quality. Sophisticated technology is employed in an effort to overcome these organizational shortcomings. Moreover, many Western suppliers of Advanced Manufacturing Technologies have traditionally favoured complex and expensive 'state-of-the-art' systems. Consequently users have found the technologies difficult to justify financially, or have subsequently experienced difficulties. In contrast, the Japanese begin with the advantage of a highly trained, multiskilled workforce, good communication between different functions and close relationships with suppliers. This has allowed Japanese manufacturers to adopt less complex technology, and to focus on improving flexibility. This suggests that firms should focus on organizational change consistent with their chosen strategy before investing in Advanced Manufacturing Technologies. The widespread adoption of Advanced Manufacturing Technologies is a necessary, but not sufficient condition for success in the 1990s.

REFERENCES

Abernathy, W. J. (1978) *The Productivity Dilemma: Roadblock to Innovation in the Automobile Industry*, Johns Hopkins University Press, Baltimore.

Atkinson, J. (1984) *Flexibility, Uncertainty and Manpower Management*, Institute of Manpower Studies Report No.89, Brighton.

Ayres, R. U. and Miller, S. M. (1985) *Robotics and Flexible Manufacturing Technologies*, Noyes, New Jersey.
Bessant, J. (1991) *Managing Advanced Manufacturing Technologies*, NCC-Blackwell, Oxford.
Blois, K. J. (1985) Matching manufacturing technologies to industrial markets and strategies, *Industrial Marketing Management*, Vol. 14, pp. 43–7.
Buchanan, D. A. and Boddy, D. (1983) *Organizations in the Computer Age: Technological Imperatives and Strategic Choice*, Gower, Aldershot.
Child, J. and Mansfield, R. (1972) Technology, size and organisation structure, *Sociology*, Vol. 6, pp. 369–93.
Dale, B. G. and Plunkett, J. J. (1990) *Managing Quality*, Philip Allan, Hemel Hempstead.
Dale, B. G. and Tidd, J. (1991) Japanese total quality control: a study of best practice, *Proceedings of the Institute of Mechanical Engineers*, Vol. 205.
ECE (1986) *Recent Trends in Flexible Manufacturing*, United Nations, New York.
Ferdows, K., Miller, J. G., Nakane, J. and Vollmann, T. E. (1986) Evolving global manufacturing strategies: projections into the 1990s, *International Journal of Operations Management*, Vol. 6, no. 4, pp. 6–16.
Fleck, J. (1987) *Robotics: Organisation and Management*, Dept. of Business Studies Working Paper No. 87/9, University of Edinburgh.
Goldhar, J. D. and Jelinek, M. (1983) Plan for Economies of Scope, Harvard Business Review, Vol. 61, no. 6, pp. 141–48.
Hayes, R. H. and Wheelwright, S. C. (1984) *Restoring Our Competitive Edge: Competing Through Manufacturing*, John Wiley, New York.
Hill, T. J. (1980) Manufacturing implications in determining corporate policy, *International Journal of Operations Management*, Vol. 1, no. 1, pp. 3–11.
Hill, T. J. (1985) *Manufacturing Strategy*, Macmillan, London.
Hounshell, D. A. (1984) *From the American System to Mass Production 1800–1932*, Johns Hopkins University Press, Baltimore.
Ingersoll Engineers (1984) *The FMS Report*, IFS, Bedford.
Ishitani, H. and Kaya, Y. (1989) Robotisation in Japanese manufacturing industry, *Technological Forecasting and Social Change*, Vol. 35, no. 2, pp. 97–132.
Jaikumar, R. (1986) Postindustrial manufacturing, *Harvard Business Review*, Vol. 6, pp. 69–76.
Kaplinsky, R. (1984) *Automation, the Technology and Society*, Longman, Essex.
Luchs, B. (1990) Quality as a strategic weapon, *European Business Journal*, Vol. 2, no. 4, pp. 34–47.
Meredith, J. R. (1987) The strategic advantages of new manufacturing technologies, *Strategic Management Review*, Vol. 8, no. 3, pp. 249–58.
NEDC (1984) *Flexible Machining Systems*, National Economic Development Council, London.
Porter, M. J. (1980) *Competitive Strategy*, Free Press, New York.
Schonberger, R. J. (1987) Frugal manufacturing, *Harvard Business Review*, Vol. 87, no. 5, pp. 95–100.
Shetty, Y. K. and Buehler, V. M. (1991) *The Quest for Competitiveness: Lessons from America's Productivity and Quality Leaders*, Quorum Books, New York.
Tidd, J. (1990) The flexible factory: robots are not enough, *Scientific American*, Scientific European supplement, October.

Tidd, J. (1991) *Flexible Manufacturing Technologies and International Competitiveness*, Pinter, London.
Watanabe, S. (1987) *Microelectronics, Automation and Employment in the Automobile Industry*, Wiley, Chichester.
Wickens, P. (1987) *The Road to Nissan*, Macmillan, Basingstoke.
Womack, J. P., Jones, D. T. and Roos, D. (1990) *The Machine That Changed The World*, Rawson Associates, New York.
Woodward, J. (1965) *Industrial Organisation: Theory and Practice*, Oxford University Press, Oxford.

3

THE PROBLEM OF IMPLEMENTING MANUFACTURING STRATEGY

Nicholas Kinnie and Roy Staughton

INTRODUCTION

Much of the interest in Manufacturing Strategy has concentrated on specific techniques such as Just-in-Time (JIT) and Materials Requirement Planning (MRP). Clearly, these techniques are of major importance and are considered in detail elsewhere in this book. There is, however, a growing body of evidence pointing to the high failure rate of many of these techniques. Often explanations for the inability to exploit the benefits of these new techniques have concentrated on technical factors. The focus of this chapter is, however, on the contribution made by the process of implementation of Manufacturing Strategy to the successful introduction of these techniques. This process, which we define as the decision-making steps followed by managers when designing and putting into practice new manufacturing techniques, has attracted relatively limited interest. Indeed, a group of influential US researchers has recently commented that 'there has been a lack of theoretical attention to the processes of formulating and implementing a Manufacturing Strategy' (Marucheck, Pannesi and Anderson, 1990). Particular attention is paid here to developing a cross-functional perspective on this process which integrates the Manufacturing Strategy, Human Resources and Organizational Behaviour perspectives.

The failure of many of the changes in Manufacturing Strategy to achieve their objectives is now well documented. Research suggests, for example, that up to 50% of companies fail to achieve their objectives with MRP implementation (White *et al.*, 1982), while another study (Archer, 1990) placed the failure rate as high as 70%. Furthermore, Wight (1990) reports that approximately 95% of companies fail to achieve the Class 'A' status they are seeking.

This chapter argues that the cross-functional perspective put forward gives greater insights into the reasons for the failure to exploit the benefits of new technology than have so far been offered. It attributes the poor performance experienced by many organizations to the ways managers

Perspective	1970	1980	1990
Manufacturing Strategy Content	■1969(Skinner)	■1984(Hayes and Wheelwright) ■1985(Hill)	
Dedicated Implementation		■1986(Callerman and Heyl) ■1988(Vollman)	■1990(Harber)
Generic Implementation			■1990(Voss) ■1990(Barker)
Human Resources and Organisational Behaviour		■1985(Senker and Beasley) ■1986(Goodridge) ■1987(Rothwell) ■1989(Oliver and Wilkinson)	■1990(Wilkinson)
Cross-Functional			■1990(Marucheck) ■1991(Akkerman and Aken) ■1991(Beatty) ■1991(Boer and Krabbendam) ■1991(Whittle)

Figure 3.1 Development of thinking in the implementation of Manufacturing Strategy

typically introduce and manage the process of implementing a Manufacturing Strategy.

Three main propositions are put forward in this chapter. First, greater attention needs to be paid to the process of implementing a Manufacturing Strategy. Second, Human Resources and Organizational Behaviour issues are central to this process of implementation. Finally, these issues are more likely to be handled successfully if they are considered throughout the design and introduction of a new Manufacturing Strategy, rather than being dealt with when the strategy is put into practice.

Much of the support for these propositions is contained in the third section of this chapter which looks at case study evidence collected by the authors. The fourth section discusses the implications of these findings, while the final section looks at areas for future research. The second section, which follows, carries out an extensive review of the development of thinking in this field, which is summarized in Figure 3.1.

PREVIOUS RESEARCH

This section describes the previous research into the implementation of Manufacturing Strategy and, in particular, traces the emergence of a cross-functional perspective.

Three perspectives on the implementation of Manufacturing Strategy are recognized by the authors: first, the content of a Manufacturing Strategy itself; second, the perspective which examines the Human Resources and Organizational Behaviour aspects of Manufacturing Strategy; and finally, the cross-functional perspective which seeks to integrate the Manufacturing Strategy and Human Resources and Organizational Behaviour perspectives. Each of these is now examined in turn.

Manufacturing Strategy Perspective

The research under the Manufacturing Strategy perspective is divided into three sub-sections: first, research dealing with Manufacturing Strategy content; second, studies of the implementation issues dedicated to particular manufacturing techniques, such as MRP and JIT; finally, the examination of generic implementation issues unrelated to any particular technique.

Manufacturing Strategy content

The early work in the field concentrated on the objectives of a Manufacturing Strategy, and put forward a notional framework for constructing one (Hill, 1985). Little attention was paid to how the strategy might be put into practice or to the Human Resources and Organizational Behaviour issues.

A number of distinctive contributions of this literature can be recognized. First, the strategic role for manufacturing within the organization was identified in the early work by Skinner (1969), who argued that manufacturing cannot perform all tasks (for example price, flexibility, quality and delivery speed) equally well, and therefore organizations have to recognize a trade-off in the way they structure their factories. Second, great stress was laid on achieving an internal consistency or 'fit' between the various elements of the Manufacturing Strategy. In particular, this involved bringing the strategy for the manufacturing function in line with the marketing strategy. Third, the key device for ensuring this consistency was the establishment of a set of 'order-winning criteria'. According to Hill (1985), the choice of manufacturing process affects the function's ability to match these order-winning criteria. The idea of a trade-off between, for example, cost and quality, has been challenged (Schonberger, 1986). Indeed, Schonberger argues that with the emergence of 'World Class Manufacturing' and Total Quality Control (TQC) it is possible to combine, in the long run, short lead times, low cost and high quality within highly flexible operations. Other authors such as Skinner (1969) and Slack (1991: 10) have recognized that the degree of trade-off can be challenged with manufacturing techniques such as Just-in-Time and Total Quality Management (TQM), although they cannot be totally eliminated.

Some limited attention was paid to implementation by authors such as Hill (1985) under the heading of 'infrastructure'. This included changes which were to be made to functional support, manufacturing systems, controls and procedures, and work structuring.

This approach is typified by the emphasis it places on the choice of particular techniques; implementation is seen as something which is dealt with after the required technique has been chosen. For example, Aggarwal (1985: 16) portrays the problem of change as 'choosing a system and making it work'. He then goes on to discuss some of the obstacles to implementation, including the existence of informal working practices and 'getting employees to perform appropriately'. The decision facing managers wishing to develop a Manufacturing Strategy is simply one of which technique they choose to run their factories.

Reviews of the early literature began to express some dissatisfaction with this emphasis on content. Anderson, Cleveland and Schroeder (1989: 2) argued, 'perhaps even more revealing in the literature is the lack of attention to the process of operations strategy. There seems to be an implicit assumption that . . . the process of operations strategy is merely perfunctory or self-fulfilling, both in an analytical and an organisational sense.'

Later research showed the beginnings of the recognition of the importance of the implementation of Manufacturing Strategy, perhaps if only because of the low rates of success of some of the more popular techniques. This led to discussions of implementation explicitly linked to particular manufacturing techniques.

Dedicated implementation

Much of the discussion of dedicated implementation has been on MRP and JIT. For example, Callerman and Heyl (1986: 31) argued that 'the problems with MRP implementation are people-related, not technical in nature'. They put forward a 'theoretical model for implementation' which 'has at its heart the relationship between people and the technical system'. The key elements of this model were education, support from management and users and involvement of management information systems staff (Callerman and Heyl, 1986: 33–4). Particular attention was also paid to the feedback system which provided reinforcement to managers and users. Vollman, Berry and Whybark (1988) also placed emphasis on the importance of implementation when changing manufacturing planning and control systems. Indeed, they argued that 'it is almost impossible to over-emphasise the role of education in achieving Class A system implementation' (Vollman *et al.*, 1988: 475).

Harber *et al.* (1990: 24) looked for 'a suitable infrastructure to implement the full-scale JIT approach'. They identified top management support, education and training, quality and relationships with suppliers as key influences. However, the place of implementation within the development of Manufacturing Strategy is clearly secondary: 'in the event that a decision is made to proceed with a JIT programme, the implementation will need careful consideration and planning' (Harber *et al.*, 1990: 28).

Following this work, some authors began to consider the implementation issues which were independent of the particular technique being considered.

Generic implementation

The more generic approach to implementation began to raise process issues but made little attempt to consider the Human Resources and Organizational Behaviour issues involved. In these studies greater emphasis was placed on the process of implementing the strategy, rather than the content of the strategy. However, much of this work was within the existing framework of Manufacturing Strategy, which saw implementation as something which took place after the strategy had been set.

For example, Voss (1990) put forward a three-phase model of Manufacturing Strategy development. This model sought to develop the Hill model by placing greater emphasis on the implementation of Manufacturing Strategy by referring to the 'process of Manufacturing Strategy'. The first phase of this model was 'aimed at developing a common understanding within manufacturing and agreement with marketing as to what the mission or task of manufacturing is'. The second phase involved ensuring 'that there is a means of making sure that the process and infrastructure decisions are consistent with the manufacturing tasks, and that they are consistent with each other'. The final phase in this model was implementation (Voss, 1990: 956–7). Although this model is useful since it recognizes implementation as an important issue it does not give any detailed attention to Human Resources and Organizational Behaviour issues.

Barker put forward two models, one which dealt with the formulation of Manufacturing Strategy, and the second which dealt with implementation (Barker, 1990; Barker and Powell, 1989). The model of formulation adopted a systems approach, and identified four sub-systems which were important: pre-production control, concurrent production control, post-production control, and throughput time (Barker and Powell, 1989: 2,045–9). The implementation model put forward an 'input adaptive strategy' which had four stages: formulation, analysis, application and refinement (Barker, 1990: 675). Some attention was given to 'human factors' in this model since it was acknowledged that 'the success of the strategy will ultimately depend upon the support of people in the organisation accepting the proposals' (Barker, 1990: 678).

Human Resources and Organizational Behaviour perspective

Research under this heading can be sub-divided into Human Resources and Organizational Behaviour.

Human Resources

Much of this research has highlighted the Human Resources issues which have to be taken into account when changing a Manufacturing Strategy, and has provided comment on the extent to which these issues are managed appropriately. However, issues outside the Human Resources area were not examined and no attempt was made to suggest ways of achieving the changes required. For example, Goodridge (1986) paid attention to employment and skill patterns, training, pay, industrial relations, organization structure, and management style. Other authors picked out particular issues for specialist attention. For example, Rothwell (1987: 50) stressed the importance of company employment policies, suggesting that these policies 'determine to a large extent the manner in which new technology is implemented'. Senker and Beasley (1985: 56) dealt with the consequences for skills when computerized production planning and control systems are introduced, arguing that the use of these systems 'involves the need for widespread changes in job and skill content'.

Further work provided some insights into the extent to which these Human Resource Management issues are actually dealt with in practice. For example, Oliver and Wilkinson compared the use of personnel and industrial relations practices by British and Japanese companies. The most striking point about their findings was 'the lack of association between the use of Japanese-style manufacturing practices and the use of Japanese-style personnel practices' (Oliver and Wilkinson, 1989: 86).

Wilkinson, Allen and Snape (1990) provided practical evidence of the possible inter-connections between a strategy for Human Resource Management and one particular change, the introduction of Total Quality Management. On the basis of observations in two organizations they found that 'TQM appears to be consistent with a move towards human resource management, not only in the emphasis on employee commitment rather than compliance, and in the underlying unitarist philosophy, but also in that both identify line managers as having a key responsibility for the management of people'. These changes therefore can be seen 'as requiring a strategic approach to the management of labour' (Wilkinson *et al.*, 1990: 31).

Similar evidence supporting this approach to change was provided by the Pirelli General case which saw the setting up of a greenfield site with a series of innovative manufacturing practices including Computer-Integrated Manufacturing, Just-in-Time production and Flexible Manufacturing Systems. Here planning began some two years before the plant was opened and the 'personnel department was involved at a very early stage in planning the introduction of new technology and devising an appropriate personnel strategy for a new automated factory' (Yeandle and Clark, 1989: 51). This meant that the company was able to develop 'an integrated and proactive set of personnel policies well in advance of the project implementation' (1989: 52).

Organizational Behaviour

Research has been carried out in the Organizational Behaviour field considering the implications of a change in Manufacturing Strategy. However, no attempt was made to put forward a model of change and little consideration is given to the practical issues of change which were involved. Oliver and Davies (1990) looked at the consequences for the organizational culture of a change in Manufacturing Strategy. They noted that in one plant the 'new Manufacturing Strategy contained one set of assumptions about how the world "ought" to be, the existing culture contained quite another', and they conclude that these new techniques 'if implemented seriously, carry significant implications for the political and cultural maps within the organisation'.

Misterek *et al.* (1991: 539) considered the same issues from a more theoretical viewpoint, looking at the links between the achievement of World Class Manufacturing status and the characteristics of organizational culture. Following statistical analysis the authors found 'a very significant relationship between organisational culture and manufacturing strategy'. The results suggested that 'low power distance, high collectivism and high congruity of the culture are associated with a significant contribution by manufacturing to the firm's competitive position'. Other researchers questioned 'how significant the cultural change must be to allow strategy implementation' (Marucheck *et al.*, 1990: 121). These researchers suggested that it may be possible 'to develop a climate for effective manufacturing strategy as a distinct sub-system of the organisational culture'. They argued that these critical sub-systems included a cost-accounting system, information system, appropriate decision-making processes, personnel performance evaluation, and compensation programmes.

Similarly, other researchers placed emphasis on the redesign of the structure of the organization. Tranfield and his colleagues (1990: 866–7) surveyed a series of changes in Manufacturing Strategy and predicted that these would result in organizational redesign in areas such as skills, work organization, functional integration, control, inter-organizational relationships and culture.

As an extension of this interest in Human Resources and Organizational Behaviour issues other researchers have conducted studies which have adopted what is termed the cross-functional perspective.

Cross-functional perspective

This third perspective seeks to integrate the Manufacturing Strategy, Human Resources and Organizational Behaviour perspectives. This research is distinctive because it examines Manufacturing Strategy implementation and includes consideration of both Human Resources and Organizational Behaviour issues. The focus is not only on the content of Manufacturing Strategy but also on the key issues involved in the process of implementation. Indeed, these approaches stress the parallel development of both the content and the process, and pay attention to the strategic and cultural changes involved.

Whittle and her colleagues (1991) provided an introduction to this type of research because they argued that too much attention has been given to the task of changing a Manufacturing Strategy, and insufficient attention given to the ways that these changes might be achieved. Tranfield *et al.* (1990: 864) argued that the key to success may well be to implement a parallel process of organizational change to go alongside the new techniques which were being introduced. The advantages of this approach are clearly articulated by Beatty (1991: 192) who argued that by 'addressing implementation issues proactively as a part of the planning for Advanced Manufacturing Technology, a coherent set of implementation choices can help start an AMT project on the road to success'.

Research in this area sought to identify firstly the issues which are critical to success and then the different ways in which these issues might be handled.

Marucheck and her colleagues made two key points on the basis of observations in six firms implementing a Manufacturing Strategy. First, they observed that the successful organizations gained most of their benefits because of the attention they paid to the implementation of Manufacturing Strategy. Second, they noted that 'implementation is a less structured and more behaviourally oriented process than strategy formulation' (Marucheck *et al.*, 1990: 121). They went on to pick out various activities which they felt were important, including the involvement of lower levels of management, and the extent of cultural change which is involved. They also pointed out that the existing cost-accounting and capital justification procedures can provide major obstacles to the implementation of the Manufacturing Strategy. Perhaps most importantly of all they observed that corporate management, rather than manufacturing executives, provided the impetus for the successful introduction of a Manufacturing Strategy.

Similarly, Boer and Krabbendam (1991) identified three groups of activities which they regard as central to what they saw as the 'manufacturing innovation process': problem-solving, internal diffusion and organizational change. They then highlighted the implementation roles which were neglected during the changes and considered their impact on the success of the whole process.

The second area of research within this cross-functional perspective was devoted to the various ways in which the change process itself might be carried out. Akkermans and van Aken (1991), for example, suggested that adopting a new perspective from the business strategy literature can help to overcome some of the problems which they perceive when implementing a change in Manufacturing Strategy. They suggested the organizational problems could be overcome by seeing strategy as politics, the problems of knowledge by seeing strategy as involving cultural change, and the problems of attitude by stressing that strategy involved organizational learning. These issues they regarded as those which 'constitute the major challenge in developing a successful operations strategy'.

Another approach was to distinguish between various models of change (Whittle *et al.*, 1991). Two of these, 'Total Institutionalization' and 'Adaptive Survival', were described as 'building palaces' because they involve highly rigid and bureaucratic approaches to change. While two others, the 'Learning Organization' and 'Minimalist Transformation', were more flexible approaches which were described as 'erecting tents' (Whittle *et al.*, 1991). The task, according to these authors was to strike the right balance between these two approaches.

Summary

This review of the literature has identified three perspectives on the process of implementing a Manufacturing Strategy: first, the Manufacturing Strategy perspective including dedicated and generic implementation; second, the Human Resources and Organizational Behaviour perspective; and third, the cross-functional perspective concerned with models of change. This work on models of change is still in its early stages; much work is needed to validate, substantiate and develop these ideas further. The following section

	Products	Process characteristics Capital/labour intensive	Product range	Batch size	Number of employees	Unionization
Components	Plastic components	Labour	Narrow	Large	600	Yes
Discs	Computer equipment	Capital	Narrow	Large	900	No
Engines	Vehicle components	Capital	Narrow	Small	150	No
Refrigeration	Refrigeration components	Labour	Wide	Large	2,100	Yes

Figure 3.2 Details of the companies studied

of this chapter provides some examples of this cross-functional approach by reference to recent empirical research.

APPROACHES TO THE IMPLEMENTATION OF MANUFACTURING STRATEGY

This section develops and applies the cross-functional perspective described in the previous section by reference to practical case studies. It provides support for the propositions put forward at the beginning of the chapter concerning the importance of the process of implementation. In particular, it has three objectives: first, to identify three different approaches to the implementation process and describe these based on the way Human Resources and Organizational Behaviour issues are handled; second, to consider why these different approaches are adopted and under what circumstances they might be appropriately used; and finally, to describe the advantages and drawbacks of each approach, particularly in terms of the impact of the approach on the outcome of the change in Manufacturing Strategy.

Much of the discussion is based upon four case studies of companies undergoing a change in their strategy for manufacturing. Details of these cases are given in Figure 3.2. Semi-structured interviews were conducted with directors and managers in all functions and data were collected over several months to provide a longitudinal picture of the changes taking place.

Three different approaches to handling Human Resources and Organizational Behaviour issues during the implementation of Manufacturing Strategy are identified:

- Wait and see;
- Learn as You Go;
- Predict and Pre-empt.

The first type, 'Wait and See', adopts a largely reactive approach; the second type, 'Learn as You Go', demonstrates an incremental approach; and the final type, 'Predict and Pre-empt', adopts a proactive approach to these issues. Each of these approaches is now discussed in turn and is summarized in Figure 3.3.

'Wait and See'

The first approach, 'Wait and See', involves dealing with implementation issues in a fire-fighting, *ad hoc* manner, handling problems when they emerge. No attempt is made to predict, foresee or plan for Human Resources and Organizational Behaviour issues. These issues are dealt with after the change in Manufacturing Strategy has been made. It is almost as if the change were simply 'plugged in', as if it were a machine, and then left to its own devices. If a Human Resources issue emerges, such as a shortfall of

Type	Characteristics	Human Resource Management issues			Examples
		Education and training	Pay systems and structures	Staffing	
'Wait and See'	Reactive, *ad hoc* solutions to specific problems	Limited on- and off-the-job training; some educational programmes	Limited attention to pay structures; some changes to pay systems	Limited attention to staffing; recruitment and retention problems	Refrigeration Components
'Learn as You Go'	Changing attitudes via feedback and learning	Practical demonstrations to educate; some job change training	Piecemeal changes to pay systems and structures	Attention to the selection and development of employees	Engines
'Predict and Pre-empt'	Strategy for HRM; change is the norm	Extensive training and education programmes by specialists and line managers	Pay systems and structures support and encourage change	Detailed attention to selection and placement of employees	Discs

Figure 3.3 Types of approaches to handling Human Resources and Organizational Behaviour issues during the implementation of Manufacturing Strategy

skills and knowledge then managers, at best, will react by arranging the necessary training. At worst, nothing will happen at all. This type of approach has attractions in that, in the short term at least, it involves no, apparently 'unnecessary', expenditure, such as that on training. Generally, however, it is this kind of approach which is associated with a failure to achieve the objectives of change. Commonly, these organizations do not have the appropriately trained staff or the pay systems to allow the benefits of the change in Manufacturing Strategy to be realized.

Two of the companies studied, 'Refrigeration' and 'Components', were typical of this approach. They paid little attention to the process or the content of their implementation of Manufacturing Strategy. There was little or no attempt to plan for the change, the ownership of the changes was very limited, and there was a strong tendency to evaluate the changes against financial criteria alone. Managers in both companies perceived the changes in a very narrow way, as largely technical rather than to do with people. Both companies failed to achieve their objectives, and over-ran their allotted time. The system installed was either used only in part or quickly fell into disrepute.

At 'Refrigeration', where MRP was introduced, many of the old manual systems remained in use long after the new computer-based system had been installed. The result was therefore something of a 'hotch potch' with the new system being used occasionally, but more commonly operators fell back on the old system. Often the users did not even have the basic skills required to operate the system and did not understand why the new system was being introduced. Largely as a result of this the performance of the MRP system was very poor and the anticipated benefits of reductions in Work in Progress (WIP) and finished goods stocks did not materialize.

The process of introducing the change involved virtually no planning, with little consideration of the likely Human Resources implications. In particular, the process was characterized by a lack of ownership of the system, since it was originally introduced by the manufacturing services director, but was then transferred to the responsibility of the finance director. Consequently, the feeling persisted amongst the employees that 'the system was dumped on us'. The change itself was perceived in very narrow terms, concerned with improving output, and, according to one planning manager, 'Manufacturing was seen as a problem, it needed sorting out.'

This attitude towards the process of change was reflected in the actions taken to deal with the Human Resources aspects of the change. Only a minimal amount of training was carried out, concerned mostly with technical issues. No attempt was made to deal with the education of the wider workforce or the possible impacts of the changes. Indeed, the company chose not to use the training schemes offered by the company providing the software or to employ an in-house trainer because of the costs involved. Following the launch virtually no supporting training was carried out, perhaps because of the change of 'ownership' which took place, and problems were dealt with solely on a 'trouble-shooting' basis.

It was eventually realized, rather late in the day, that there were Human Resources implications of this change. For example, when it came to recruiting new planners it was soon evident that, in the words of one manager, 'a very different type of "animal" was required'. Whereas previously, the most highly skilled shopfloor employee might be promoted to the post of planner, managers now wanted to recruit planners with a knowledge of computer systems and shopfloor operations. In practice, they recruited a mix of new graduates and experienced skilled employees.

A very similar approach was adopted at 'Components' where two major changes, an MRP system and a move from make-to-stock to make-to-order, were introduced virtually simultaneously. JIT supply was seen as vital to respond to market demand, but the MRP system was, as one manager said, 'simply unable to produce what was required of it'. The system had 'blundered along since implementation', being used only for the generation of orders. It required frequent manual intervention and a high level of maintenance. Indeed, one planning manager said, 'Some of our foreman have gone back to paper and pencil.' The time-scale was seriously over-run, and the intended changes were not completed. However, eventually the company 'solved' its problem of becoming a JIT supplier by holding very high finished goods stocks to meet variations in demand.

The process of implementation paid virtually no attention to planning, despite 'Components' being based on a 'greenfield' site. According to one manager, 'We just set up the plant, we didn't think how we were going to plan or control it.' Indeed, the MRP system was put in without any real analysis of what its purpose was to be. One senior manager said, 'We were sold a solution to a problem we didn't have . . . we were buying what someone was selling.' The ownership of the system was not widespread, and most users adopted the 'not made here' view when a problem arose. In fact, first-line managers took great delight in contacting the IT manager, who was based at another site, to inform him that 'Your system has gone down again.'

This attitude towards the system was partly caused by the absence of an effective training and education programme. This was particularly noticeable because of the other changes which were taking place at the same time. The introduction of the MRP system with a JIT philosophy meant that employees were being asked to make two changes without the effects of either of them being explained. Consequently, one manager claimed that 'the figures from the new system appeared to be quite crazy and supervisors and line managers went back to their hand-written calculations'. This problem was made worse by the lack of data-entry discipline, because employees did not appreciate the significance of the data they were inputting.

'Learn as You Go'

The second approach, 'Learn as You Go', seeks to develop rules of good practice incrementally on the basis of experience. Companies adopting this approach make changes in a piecemeal fashion to create a culture or climate which is amenable to change. However, there is an absence of any kind of thought-through strategy towards Human Resources, although managers actively seek to learn lessons from their experience of introducing change and build up rules as they go along. Particular attention is paid to good communications and feedback mechanisms to allow a consistent approach towards managing change to be developed gradually. This approach has a number of attractions, particularly to organizations which are growing quickly or are operating in a fast-changing environment. The approach is flexible and adaptive and does not consume large amounts of specialists' time in developing a strategy. However, the potential drawback is that it is highly dependent on the efficiency of the communications and feedback mechanism. If this mechanism fails to operate, the process of implementation could become purely reactive, as seen in the 'Wait and See' approach.

'Engines' provides a good example of this approach. This organization sought to introduce JIT and its experience could be described as being in a transitionary stage. JIT was not applied to the whole organization, but on a pilot basis to a single department. Experience was generally good, although the time-scale was over-run considerably. However, the changes led to a large reduction in stock from nine weeks to four to five days. After some time though there were problems with quality and with breakdowns of machinery, both of which led to interruptions to supply. Consequently, the level of stock and Work in Progress (WIP) remained higher than necessary. However, there were none of the problems of staffing, payment and training which were seen in some of the other companies.

This experience can be attributed to a number of influences. First, whilst not deliberately taking a strategic view, managers realized that a long-term perspective was required. They sought to develop, over time, attitudes which were amenable to change. Indeed, the company had ambitious plans to expand its operation on to a greenfield site and was to some extent using the present site as a pilot for the future. The ownership of changes in 'Engines' was widespread, although they were driven by a small group of senior managers who had wide experience of other more reactive approaches. Evaluation of the outcome was not restricted simply to the financial benefits, but also took into account the wider changes in attitudes.

A good example of the commitment to the changes of this kind is given by the way in which the concept of JIT was introduced to the workforce. Senior managers staged a practical demonstration of JIT using cardboard boxes and spacers to simulate their own assembly line and to illustrate the differences between 'push' and 'pull' manufacture. This proved to be a very persuasive tool for introducing employees to the principal benefits of JIT.

Similar attention was also paid to the content of the change programmes. Training programmes were carefully designed to cope with the role changes that supervisors, for example, experienced with the adoption of JIT. Supervisors were not being asked to take on a team leader role, and indeed, in the words of one manager, 'supervision has taken on a whole new meaning'. It was no longer appropriate for supervisors to adopt an authoritarian approach when employees were being asked to act in a creative and flexible way. Existing supervisors needed to be trained in the motivation, communication and appraisal techniques which were now integral to their new jobs.

Steps were also taken to ensure that the payment system was appropriate to the new Manufacturing Strategy. Managers deliberately chose to avoid an individual bonus scheme because they wanted to do away with the 'keep the machine running – pile 'em high' approach to manufacturing. Instead, they wanted to promote a quality-driven approach based on flexibility and teamworking. Overtime was not paid for, although employees had a guaranteed annual payment as part of the overall package known as the Stable Income Plan.

New operators and supervisors were selected very carefully at 'Engines' so that they would be comfortable with the company culture and philosophy. Employees were sought who were willing to take on responsibility, prepared to accept change, and likely to identify strongly with the style of the site. Indeed, senior managers were reluctant to transfer employees from other parts of the group to which 'Engines' belonged because they might be familiar with a different set of working practices, and their strong preference was to recruit employees who were new to the company. The final element in the approach to staffing was to ensure that no detailed job descriptions existed and that employees were multiskilled so that they were able to operate very flexibly.

'Predict and Pre-empt'

The third approach, 'Predict and Pre-empt', places emphasis on the process of implementation and the content of changes involved. Often this involves adopting a series of principles which views change as the norm rather than the exception. The 'Predict and Pre-empt' approach involves taking account of Human Resources and Organizational Behaviour issues at the outset of the planning process so that these issues are considered alongside or even ahead of the technical concerns. This approach is clearly appropriate to an organization which already adopts a strategic approach to managing its business and has the necessary staff resources to permit this. However, it must be recognized that this kind of approach takes some considerable time to develop and cannot be put in place 'overnight'.

'Discs' provided a good example of this approach when it was introducing a new product and production facility. These changes were made successfully and the new product was introduced on time. There were none of the unforeseen problems of the kind seen in the other cases. Operatives on the line had the correct skills and knowledge for the product, and their superiors knew what targets they were expected to deliver. The output of the line increased as expected over several months. These changes were seen as unexceptional, as part of a continuous programme where change was commonplace. This success can be attributed to a number of factors characteristic of the approach adopted.

Perhaps most importantly the process of introducing change was seen to be part of the company's overall approach towards managing Human Resources. The aim was to provide a series of facilitating measures covering education, training, payment and staffing which would ease the introduction of any change in Manufacturing Strategy. A good example of this proactive approach was provided by the appointment of a training specialist whose job was designed to fill the gap between the 'hands-on' training provided for operators, and the more formal general courses provided by the personnel department. This position required good knowledge not only of the production processes involved, but also of the design and implementation of training programmes. Once production started, this person was also able to provide a training diagnostic service, seeking to identify ways in which employee performance could be improved via training.

The extent of the changes involved was perceived to take into account the non-technical as well as the technical considerations. Extensive education and training programmes of both a general and specialist nature were provided for all employees and dealt with manufacturing issues as well as all aspects of company strategy towards Human Resources. The education of the workforce was not solely in the hands of the specialists since the established communication channels were also used to reinforce the messages. Supervisors, for example, had the responsibility of communicating the benefits of the established JIT philosophy, virtually on a day-to-day basis. According to one supervisor, 'It's my job to explain why, sometimes, there are no components, and how this fits in with JIT.'

The payment system at 'Discs' sought to encourage flexibility and a willingness to change. There were no annual increases in pay, instead employees were rewarded on the basis of the evaluation of their performance made by the supervisor. Similarly, the approach to staffing provided a good example of the interface between the management of Human Resources and changes in Manufacturing Strategy. The stated strategy towards Human Resources was based on high wages, good benefits and guaranteed employment. Conditions in the marketplace were, however, volatile, bringing some quite major shifts in the demand for, and mix of products. 'Discs' coped with this by employing a high proportion of temporary employees (approximately 25% of the workforce) who did not enjoy the same guarantee of employment as the permanent workforce.

Ownership of changes in 'Discs' was also very widespread, reflecting the company philosophy to encourage the involvement of employees at all levels. For example, the strategy towards Human Resources recognized that these issues were the responsibility of all managers both at a senior and a junior level. This feeling of ownership was also enhanced by the fact that many of the employees had been with the plant since it was established around ten years earlier. Therefore, in the words of one manager, 'Many of our systems were taken for granted and came with the walls and the roof.'

Finally, the benefits of the approach were broadly defined, being concerned not just with the financial benefits in the short run, but also the contribution made towards the overall goal of improving quality.

Summary

This section has described and analysed three types of approach to the implementation of Manufacturing Strategy. Each of these types treats the Human Resources and Organizational Behaviour issues in a different way. The following section continues this discussion by turning attention to the theoretical and practical applications of these findings.

DISCUSSION

This section first proposes, on the basis of the findings discussed above, a revised version of the theoretical model put forward by Hill (1985). Discussion then turns to the implications for practitioners.

Theoretical implications

The theoretical implications concern the content of any Manufacturing Strategy and the process by which it is implemented. The existing model of Manufacturing Strategy as put forward by Hill (1985), and shown in Figure 3.4, is unsatisfactory from both the content and process viewpoints and there

1 Corporate objectives	2 Marketing strategy	3 How do products win orders in the marketplace?	4 Manufacturing Strategy	5
			Process choice	Infrastructure
Growth	Product markets and segments	Price	Choice of alternative processes	Function support
Profit	Range	Quality	Trade-offs embodied in the process choice	Manufacturing systems
Return on investment	Mix	Delivery, speed, reliability	Role of inventory in the process configuration	Controls and procedures
Other financial measures	Volumes	Colour range		Work structuring
	Standardization versus customization	Product range		Organizational structure
	Level of innovation	Design leadership		
	Leader versus follower alternatives			

Figure 3.4 Hill model of Manufacturing Strategy

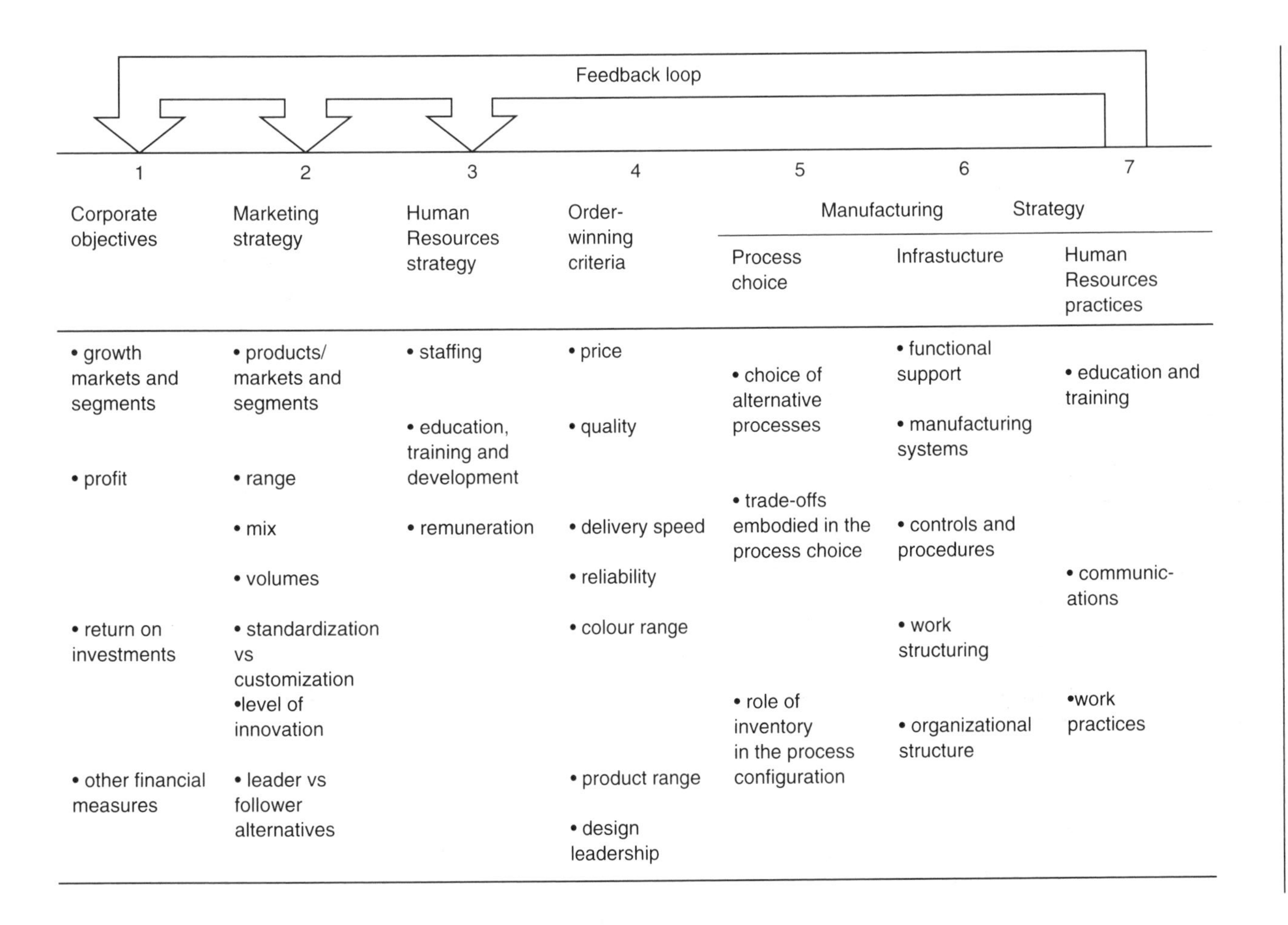

Figure 3.5 Model of the implementation of Manufacturing Strategy

are a number of possible modifications. These modifications are included in the model of the implementation of Manufacturing Strategy put forward in Figure 3.5.

First, a number of additions are required to the content of the Hill model so that the Human Resource and Organizational Behaviour issues involved in any change in strategy are thoroughly considered. The column headed 'Infrastructure' in the Hill model is inadequate and requires development. The proposal, shown in the new model of implementation in Figure 3.5, is to add an additional column to the Hill model dealing specifically with Human Resources practices including, in particular, education and training, communications and work practices. This column signifies that there are issues of this kind which need to be addressed. Typically, these issues are dealt with on a reactive basis.

Second, the findings on the process of implementation require two further modifications to the Hill model. There is a need for a further additional column, shown in the new model of implementation, which deals with Human Resources strategy. This is inserted in the Hill model between 'Marketing strategy' and 'Order-winning criteria', under a new Column 3, and includes education, training and development, staffing and remuneration. The inclusion of this column identifies the issues which require attention at an early stage if an organization is to adopt a proactive approach.

In addition, the new model of implementation includes a feedback loop between the new Column 7 and Columns 1, 2 and 3. The feedback loop reflects the opportunity for organizations to learn from the experience of changing their Manufacturing Strategy. For example, the practical implementation of a change will provide the opportunity for an organization to modify its approach in the future by learning to develop a more strategic approach on the basis of past experience (as in the case of 'Engines').

Practical implications

These findings also have a number of implications for practitioners. For example, the typology developed can serve as a comparative framework for managers considering changing their Manufacturing Strategy. First, managers are provided with a means of typifying the approach which they are currently adopting towards the implementation of Manufacturing Strategy and suggesting the consequences that this might have for the success or otherwise of the change. Second, the typology also suggests other ways in which the process of implementing Manufacturing Strategy might be handled and suggests some of the characteristics of these other approaches.

In essence, this typology provides managers with the opportunity to analyse their current approach to the implementation of Manufacturing Strategy and, in particular, their attitude to Human Resources and Organizational Behaviour issues. They might, for instance, pose the following questions when reviewing their situation: What is our current approach to implementation? What are the likely consequences of this approach for the success of Manufacturing Strategy? What alternative approach might be adopted in the future? What changes need to be made to accomplish this approach? Consideration is given elsewhere to how these changes might be achieved (Staughton et al., 1992).

CONCLUSION

This chapter has sought to advance a cross-functional perspective which focuses on the process of implementing a Manufacturing Strategy. It is argued that this perspective offers distinctive insights into the success or otherwise of the change in Manufacturing Strategy. Particular attention has been paid to the Human Resources and Organizational Behaviour aspects of these processes.

The first section of this chapter was concerned to outline the development of thinking in this field, looking at the Manufacturing Strategy, Human Resources and Organizational Behaviour and cross-functional perspectives. The second section used empirical data to illustrate the importance of this cross-functional perspective. It identified three different approaches to these processes and considered the influences on and implications of each of these. The third section outlined the theoretical and practical implications of these findings, and put forward a model to understand the process of implementing a Manufacturing Strategy.

There are a number of issues which have not been addressed by the current research and which are worthy of further attention. These are concerned first with the theoretical implications and second with the implications for practitioners.

Future research should seek to validate the proposals put forward here identifying three types of approach, and more importantly the links between the approach adopted and the outcome of the change in Manufacturing Strategy. This might take the form of either quantitative or qualitative research or a combination of the two.

Further research considering the practical implications might concentrate on how an organization might change from one approach to the process of implementing Manufacturing Strategy to another approach. Particular insights would be gained from studying the practical steps and techniques which might be followed by managers either internal or external to the organization. The most appropriate means for gathering this information is likely to be in-depth studies of organizations carried out over long periods of time.

REFERENCES

Aggarwal, S. C. (1985) MRP, JIT, OPT, FMS? Making sense of production operations systems, *Harvard Business Review*, September/October, pp. 8–16.
Akkermans, H. A. and van Aken, J. E. (1991) Problems in operations strategy development: reconciling theory with practice, in D. Bennett and C. Lewis (eds.) *Achieving Competitive Edge*, Proceedings of the 6th International Operations Management Association UK Conference, Springer Verlag, pp. 3–10.
Anderson, J. C., Cleveland, G. and Schroeder, R. (1989) Operations strategy, a literature review, *Journal of Operations Management*, Vol. 8, no. 2, pp. 1–26.
Archer, G. (1990) MRP: a review of failures and a proposal for recovery using CBS, *BPICS Control*, December.
Barker, R. C. (1990) Input adaptive strategy: formulation, development and application of a functional manufacturing strategy, *International Journal of Production Research*, Vol. 28, no. 4, pp. 675–83.
Barker, R. C. and Powell, N. K. (1989) A heuristic approach to the formulation of Manufacturing Strategy, *International Journal of Production Research*, Vol. 27, no. 2, pp. 2,041–51.
Beatty, C. A. (1991) Critical implementation decisions for advanced manufacturing technologies, in D. Bennett and C. Lewis (eds.) *Achieving Competitive Edge*, Proceedings of the 6th International Operations Management Association UK Conference, Springer Verlag, pp. 187–92.
Boer, H. and Krabbendam, K. (1991) The effective implementation and operation of flexible manufacturing systems, in D. Bennett and C. Lewis (eds.) *Achieving Competitive Edge*, Proceedings of the 6th International Operations Management Association UK Conference, Springer Verlag, pp. 64–9.
Callerman, T. E. and Heyl, J. E. (1986) A model for materials requirements planning implementation, *International Journal of Operations and Production Management*, Vol. 6, no. 5, pp. 30–7.
Goodridge, M. (1986) Operations management of human resources in the 1990s, *International Journal of Operations and Production Management*, Vol. 6, no. 4, pp. 42–60.
Harber, D., Samson, D. A., Sohal, A. S. and Wirth, A. (1990) Just-in-Time: the issue of implementation, *International Journal of Operations and Production Management*, Vol. 10, no. 1, pp. 21–30.
Hayes, S. C. and Wheelwright, S. (1984) *Restoring our Competitive Edge*, Wiley, New York.
Hill, T. J. (1985) *Manufacturing Strategy*, Macmillan, London.
Kinnie, N. (1989) Human resource management and changes in management control systems, in J. Storey (ed.) *New Perspectives on Human Resource Management*, Routledge, London, pp. 137–51.
Marucheck, A., Pannesi, R. and Anderson, C. (1990) An exploratory study of manufacturing strategy process in practice, *Journal of Operations Management*, Vol. 9, no. 1, pp. 101–23.
Misterek, C. A., Bates, K. O., Morris, W. A. and Schroeder, R. G. (1991) Manufacturing strategy and organisation culture: theory and measurement issues, in C. A. Voss (ed.) *Manufacturing Strategy – Theory and Practice*, Proceedings of the 5th Operations Management Association Annual Conference, University of Warwick, pp. 531–55.

Oliver, N. and Davies, A. (1990) Adopting Japanese style manufacturing methods: a tale of two (UK) factories, *Journal of Management Studies*, Vol. 27, no. 5, pp. 555–70.

Oliver, N. and Wilkinson, B. (1989) Japanese manufacturing techniques and personnel and industrial relations practice in Britain: evidence and implications, *British Journal of Industrial Relations*, Vol. 27, no. 2, pp. 73–91.

Rothwell, S. (1987) Selection and training for AMT, in T. Wall, C. Clegg and N. Kemp. *The Human Side of Manufacturing Technology*, John Wiley, Chichester.

Schonberger, R. (1986) *World Class Manufacturing: The Lessons of Simplicity Applied*, Macmillan, London.

Senker, P. and Beasley, M. (1985) Computerised production inventory control systems: some skill and employment implications, *Industrial Relations Journal*, Vol. 16, no. 3, pp. 52–7.

Skinner, W. (1969) Manufacturing – the missing link in corporate strategy, *Harvard Business Review*, May–June, pp. 136–45.

Slack, N. J. (1991) *The Manufacturing Advantage*, Mercury, London.

Staughton, R. V. W., Kinnie, N. J., Davies, E. H. and Smith, R. L. C. (1992) Modelling the manufacturing strategy process, *Operations Management Review*, Vol. 9, no. 2, pp. 48–68.

Tranfield, D., Smith, S., Ley, C., Bessant, J. and Levy, P. (1990) Changing organisational design and practices for computer integrated technologies, in C. A. Voss (ed.) *Manufacturing Strategy: Theory and Practice*, Proceedings of the 5th Operations Management Association Conference, University of Warwick, pp. 862–73.

Vollman, T. E., Berry, W. L. and Whybark, D. C. (1988) *Manufacturing Planning and Control Systems* (2nd edn), Irwin, Homewood, Illinois.

Voss, C. A. (1990a) *Manufacturing Strategy: Theory and Practice*, Proceedings of 5th Operations Management Association UK Conference, University of Warwick.

Voss, C. A. (1990b) The process of manufacturing strategy implementation, in C. A. Voss (ed.) *Manufacturing Strategy, Theory and Practice*, Proceedings of 5th Operations Management Association UK Conference, University of Warwick, pp. 949–62.

White, E. M., Anderson, J. C., Schroeder, R. G. and Tupy, S. E. (1982) A study of the MRP implementation process, *Journal of Operations Management*, Vol. 2, no. 3.

Whittle, S., Smith, S., Tranfield, D. and Foster, M. (1991) Implementing total quality: erecting tents or building palaces, in D. Bennett and C. Lewis (eds.) *Achieving Competitive Edge*, Proceedings of the 6th International Operations Management Association UK Conference, Springer Verlag.

Wight, O. (1990) *Survey Results: MRP/MRPII, Just-in-Time*, The Oliver Wight Companies, Essex Junction.

Wilkinson, A., Allen, P. and Snape, E. (1990) TQM and the management of labour, *Employee Relations*, Vol. 13, no. 1, pp. 24–31.

Yeandle, D. and Clark, J. (1989) Personnel strategy in an automated plant, *Personnel Management*, June, pp. 51–5.

4

THE DESIGN-MANUFACTURING INTERFACE

Arthur Francis

INTRODUCTION

As Kim Clark and colleagues observe (Clark, Chew and Fujimoto, 1992: 201), 'design and manufacturing tasks have traditionally been thought of as distinct, radically different activities'. There has been a wall between them. Over this wall the designers have thrown their drawings and specifications and production engineers have taken a pride in their ability to work out how to manufacture anything the designers could throw at them.

This dissociation is not in the spirit of New Wave Manufacturing. As companies attempt to improve the quality of their products, their speed to market with new products, and the economy with which they can do all this, close links between design and manufacture and Design *for* Manufacture (DFM) are of sharply increasing importance (Womack, Jones and Roos, 1990; Clark and Fujimoto, 1991).

In this chapter we rehearse the reasons why design and manufacturing must come closer together, look at the mechanisms by which this can be done and the improvement in strategic capability that such a move can make, and end by discussing the difficulties facing Western firms in making the appropriate changes to the architecture and culture of their organizations.

WHY THE DESIGN–MANUFACTURING INTERFACE IS CRITICAL TO COMPETITIVE SUCCESS

In many markets companies are now having to compete on at least four fronts. Not only must their products be competitive in price and quality, increasingly they must also be able to offer short lead times from initial concept to customer use and to offer an adequate number of product variants to meet specific market segments. Fierce competition across these four fronts is evident in many industrial sectors – turbine generators for power stations (Francis *et al.*, 1982), automobiles (Womack *et al.*, 1990), and consumer electronics (Sanderson, 1992), for example. Some of these products are made one-off to customer order, others are mass-produced on the basis of design requirements developed in-house. In every case the management of the relationship between design, development and manufacturing is key to maintaining a competitive position on each of these four criteria.

It hardly needs to be spelled out that the detailed design of a product can have a profound effect on the cost of its production. The construction materials, the methods of fastening it together, the way in which parts can be machined, cast or pressed are all specified by the designer and all have cost implications, but if the designer's prime concern is function in use and aesthetics, cost of production may not be given adequate attention. It needs collaboration between designer and production engineer to achieve an appropriate balance between cost and other design considerations.

Product quality can be defined in a number of ways. Clark and Fujimoto (1991) focus on two aspects of quality. The first is conformance – how well the actual product conforms to the design once it has been manufactured (how well made it is, in other words). The second is the quality of the design itself – how well the design satisfies customer expectations. Though this latter quality aspect is largely within the control of the designer alone, the conformance quality of the product relies on close collaboration between designer and manufacturer. Something which is badly designed in the sense that it is difficult to make is much less likely to have a high build quality.

Cost and quality have always been crucial competitive issues, even though firms have not always recognized that a good design–manufacturing interface can improve their competitive edge here. It is more recently that short lead time and product variety have also become competitive issues. It is worth noting here three reasons why these are now so important.

First is increased affluence, allowing more sophisticated and varied tastes to be met. This has allowed product markets to become more segmented and has created the possibility for competition on the basis of increased product variety. Henry Ford's Model T car (only available in black) and, more recently, the VW Beetle, have been superseded not just by a greater variety of different models of family car but also by a growing range of types of personal motor vehicle – small city cars, seven-seat family wagons, four-wheel-drive vehicles with an apparent potential for off-road use, for example. The wide range of different models of personal stereos on the market is another example.

Second is technical progress, with technical breakthroughs and new technologies. These allow new features to be added and new standards of performance to be achieved in existing products, and create the possibility for streams of new products. Examples of this range from the vast increase in capacity of turbine generator sets over the last few decades to the rapid increase in performance of laptop computers within the past three years.

Third is the development by Japanese companies of managerial and organizational techniques which have enabled them to improve their performance across all four competitive arenas simultaneously. In the automobile industry, for example, not only do Japanese cars have a cost and quality advantage. Japanese companies can now also do their New Product Development (NPD) quicker and more cheaply. The European and US volume car producers take, on average, about 25% longer and over 75% more engineering hours to launch a new car model than their Japanese competitiors (Clark and Fujimoto, 1991: 75, 80) and, in the case of the Americans, despite taking so much longer and spending so much more engineering effort, the quality of the new product is inferior (1991: 83). This has allowed the Japanese to build up about twice as many different model offerings on the market as their Western counterparts (Womack *et al.*, 1990: 120). In consumer electronics, Japanese firms now have a monopoly of some product areas – domestic video equipment, for example – because European and American companies can no longer compete in NPD.

Firms in many product-market segments might do better to change their competitive strategy to that now used by the Japanese car producers, suggest Susman and Dean (1989). They suggest that Western firms are now choosing between three strategies. The first may be a viable strategy for companies in mature markets making commodity-like products that typically compete on the basis of price. For these companies it may be enough to compete on only three of the four competitive fronts – low cost, high quality, and short lead time from order placement to delivery. Product variety is not important, though even for this market companies need to make effective use of DFM. However, there are increasingly few of these markets and an alternative competitive strategy is increasingly necessary. This is to compete in multiple segments. This strategy requires companies to differentiate their products by market segment and seek to earn a premium price in each segment by tailoring the product to that segment's customers. A third strategy is that of competing by continuous product improvement, frequently introducing new and improved product models and earning a premium price for them because their products are superior to those of their competitors and difficult for them to copy and sell at a lower price. All three strategies require a strengthened design–manufacturing interface.

Improving the design–manufacturing interface is in part a novel problem but in part a question of technology transfer. It is clear that in some industries, notably automobiles, the Japanese have developed a novel organizational 'technology' and that Western firms have been struggling over the last ten years to copy, or transfer, this technology. They have learned that it is now no longer adequate simply to trade off one or more of the goals of the competition – low price, high quality, short lead time, and product variety – against the others. De Meyer (1992) shows that most Western firms are far from the efficient frontier where such trade-offs become necessary. They urgently need to move closer to that frontier. But it is not just a question of mimicry. Western firms face novel problems because they are embedded in institutional arrangements that may make certain features of Japanese practice unfeasible or inappropriate. They also face a problem which the Japanese did not have – that of changing present practices to suit the new competitive strategies.

This faces Western companies with three challenges. The first is to learn about the organizational and managerial techniques which the Japanese have developed to achieve such high performance in NPD, particularly, in the context of this chapter, in the way they manage the design–manufacture interface. The second is to learn what adaptations of the Japanese method are necessary to fit a particular set of national institutional characteristics. The third is that of learning how to overcome the enormous amount of inertia within any organization that resists change of the magnitude implied by the evidence about the performance gap. The rest of this chapter focuses primarily on the first of these challenges, and says just a little about the second and third.

HOW DO THE JAPANESE DO IT?

Clark and Fujimoto's detailed international comparative study of the car industry (1991) resulted in three key findings. Japanese car manufacturers were at the leading edge in quick and efficient New Product Development because they practiced simultaneous engineering, had strong project leadership, and confronted conflicts within the design team at the beginning rather than the end of the process. We now look at each of these practices in turn.

Integrated problem-solving and simultaneous engineering

Clark and Fujimoto identify four major stages in New Product Development for cars – concept engineering, product planning, product engineering, and process engineering. Liker and Fleischer (1992: 233) suggest that most US companies are still run by machine bureaucracies with the design process organized as a set of separate specialities. They use the metaphor of factory chimneys to illustrate the way in which each function is entirely separate from the others and oriented upwards, the service provider in any one function more interested in satisfying his or her boss higher up in the chimney than serving the needs of the customer in the next chimney. It is in this kind of organization that operations are done sequentially, with the output of one function 'thrown over the wall' or, to extend the metaphor, thrown out of one chimney in the hope of landing in the next one. There are at least two problems created by this approach. The one is that there is no communication between the functions and hence no integrated problem-solving. The other is that the process is wholly sequential. Work cannot begin on the downstream process until it is completely finished upstream. This lack of communication diminishes the quality of the design, and the lack of overlapping work between stages delays its completion.

The Japanese approach, in the car industry, is to create a much more effective interface between the stages of the process. Overlapping the various stages, now increasingly known as simultaneous engineering, is the obvious first step. Vital time can be shaved from the project lead time if process engineers can begin work on an NPD project before design work is fully complete. Obviously this will work better if those in the upstream stage feed information downstream on a piece-by-piece basis rather than batching up the information and transmitting it down the line only occasionally. Moreover the earlier such transmission can start the better, though only if there is a high level of trust, understanding and lack of recrimination between the functions. If engineers fear that the early release of tentative designs mean that they will be blamed later when they have to change those designs then it is likely that they will be reluctant to co-operate in the development of simultaneous engineering. Also there is the nature of the communication itself. It should be two way and, preferably, face to face. This may seem an obvious point but there is at least one large and famous British engineering company where design changes are communicated via the internal postal service.

Matrix management

Simultaneous engineering and integrated problem-solving seem to require a particular kind of organizational architecture. Every one of the Japanese automobile companies studied by Clark and Fujimoto used some form of matrix management. In what they termed the 'heavyweight product manager' structure there was an 'overlay matrix' (Knight, 1977). Under such an arrangement the organization is still largely functional but is overlaid by another organization, based on the products under development at that time. Each product is championed by a 'heavweight' product manager whose responsibility cross-cuts the functional chief's. Engineers within each function report both to their functional head and to the relevant product manager. These product managers exercise strong direct and indirect influence across all functions and activities in their project. They are responsible not only for internal co-ordination, but also for product planning and concept development. The heavyweight project manager effectively functions as a general manager of the product.

In some Japanese firms they found what they termed a 'project execution team' structure or what Knight would term a 'secondment matrix'. In this set-up the project manager works with a team of people who devote all their time to the project, who are, in other words, seconded to the project for its duration.

There did not seem to be any difference in performance between those companies operating with heavyweight product managers and those with a project execution team but it was clear that the majority of Japanese firms used one or other of these organizational designs and that there was a continuing trend towards their increased use (Clark and Fujimoto, 1991). It would seem that some form of strong horizontal integrating mechanism across functions, to supplement the strengths of the vertically organized functions, is necessary to activate simultaneous engineering and integrated problem-solving.

Up-front confrontation of conflict

A third feature of what Womack calls 'lean design' is the up-front confrontation of conflict in design trade-offs. Clark and Fujimoto found that many Western development efforts failed to resolve critical trade-offs until very late in the project. There was a great reluctance to confront conflicts directly. Whereas in the West the number of design changes peaks at about the time the product goes into production – as difficult decisions can be delayed no longer – in Japan they are highest near the beginning of the design process. This is reflected in the number of people involved.

> In the best Japanese lean projects . . . [these] are highest at the very outset. All the relevant specialties are present, and the *shusa*'s (the large-project leader) job is to force the group to confront all the difficult trade-offs they'll have to make to agree on the project. As development proceeds, the number of people involved drops as some specialities, such as market assessment and product planning, are no longer needed.
>
> (Womack *et al.*, 1990: 115)

The typical Western case is for numbers to peak close to the time of product launch as extra bodies are brought in to resolve problems that should have been cleared up in the beginning.

Barkan (1992) reviews a number of other studies of this aspect of the design process, citing a number of other industries in which the same dramatic differences between Japanese and Western companies have been found, in particular the extent to which in Japan the number of design changes per week peaks early in the project, while work on the manufacturing systems is in its early stages. Design changes in Japanese companies have reduced almost to zero by the time they peak in typical US companies, where they peak at about the point that production is starting. Such late design changes throw an enormous strain on the relationship between design and manufacturing, especially if they are attempting simultaneous engineering, or, putting it another way, the ability to get most design changes done up-front plays a major part in making feasible simultaneous engineering. Barkan identifies a number of, apparently copyable, tools and procedures used by Japanese companies to ensure design changes take place at an early stage. These include establishing a sound definition of the project at the outset by generating a broad statement that captures the character of the product as well as the essence of the distinctive features needed to assure its success; restricting design innovations to proven technology (called 'shelf engineering' in some companies); preserving experience so that it can be reused the next time round – much easier if the organization is stable and if there is a positive constructive spirit in which comprehensive reviews of projects can be carried out; careful use of prototyping; the use of interdisciplinary teams in simultaneous engineering; and the use of structured methodologies and disciplined procedures (Barkan, 1992: 59ff.).

Other Japanese NPD techniques

In addition to their use of these organizational and managerial techniques, Japanese producers differ in a number of other respects, each of which has been claimed by one or more commentators to increase NPD efficiency and effectiveness.

One of these is the relationship between Japanese Original Equipment Manufacturers (OEM) and their suppliers. Clark and Fujimoto document the stark contrast between the typical arrangement in the US auto industry and that in Japan. Supplier–OEM relationships in the Japanese automobile industry are characterized by commitment and reciprocity and organized through three or four tiers – prime sub-assembly suppliers being supplied by component suppliers, in turn supplied by materials suppliers, etc. By contrast, US companies typically have arm's-length short-term relations with many separate companies to carry out different manufacturing steps (for example, for die-making there might in the US be moulding suppliers, casting specialists, machine shops and jig suppliers, while some Japanese companies use die suppliers who offer the die development process as a complete package).

The Japanese hierarchy of stable supply relationships is used by the OEM to contract out a substantial part of the design work. Though this contracting-out imposes a substantial co-ordination burden upon the enterprise the commitment and reciprocity in the OEM–supplier links makes this manageable, it is argued. It also allows more widespread design experience to be used and, arguably, makes easier the practice of simultaneous engineering.

'Quality Function Deployment' (QFD) is another practice, perhaps like quality management, that has been imported from Japan though it may have been invented in the USA (Rosenthal and Tatikonda, 1992). The author first heard of its use in the UK at Rover, where they had learned it from Honda. QFD is a set of techniques for improving communication between marketing product planners and product engineers. Use of QFD techniques would enable the marketing people's demand for 'a nice smooth ride' or 'a sporty feel' for a car to be translated into specific targets for vibration levels and spring rates, for example, and for this translation to be done at the design stage rather than when a prototype is on test.

Of equal importance to improvements in getting an initial product to market is the need to become quicker and more efficient in continuously upgrading and improving that product over time. Sony modularizes its designs, creating design platforms from which to spin out product variety. The multiplicity of product variants on the Walkman and their portable CD player are examples of this. There are five core designs that make up the bulk of Sony's product introductions. They have special 'tiger teams', separate from those executing product improvements and variations, who conduct aggressive technological development (Sanderson, 1992: 39). These teams are given 'clear stretch goals and longer time frames'. Their tasks have included the development of the super-small WM-20, ultralight headphones, super-flat motors, and rechargeable batteries the size and shape of a stick of chewing gum. There are also marketing/industrial design teams in Japan and in major regional markets which develop new product concepts attuned to market needs and trends. They develop product specifications and mock-ups which are then transferred to the engineers in Japan for engineering implementation, via modifications of one of the core device designs. Examples of what they have produced are the 'My First Sony' line and the rugged-style sports and 'Outback' series.

The extent to which Japanese companies use product strategies that make use of product families is not emulated by the Americans, says Sanderson, partly because they have not yet developed the methodologies for their effective deployment. There is a need to develop new computer-based techniques as well as new organizational forms. She claims that most (US) firms are not currently organized to handle the integration of large amounts of diverse and complex information, which includes not only considerations of product form, function and fabrication but also the organizational and administrative procedures that underlie the design and engineering process.

'There is a pressing need to develop a methodology for designing products, developing manufacturing processes, analysing the cost impacts of alternative designs, and integrating these decisions into the normal work environment' (Sanderson, 1992: 45). Examples of the tools are computer networks, common databases, graphical interfaces and distributed control and software. Particular design technologies are database technology, networking and communication technology, and automated design and planning tools.

Other tools include using 'virtual design' to ease the transition between product cycles. Virtual design is the process in which all the information about the product and its manufacture is captured in a virtual (computer) environment. 'Holding over' information at a high level of abstraction about previous designs is another way of smoothing the transition between product generations. CAM and CAD/CAE are technologies that can incorporate 'virtual designs' and 'hold over' information.

De Meyer (1992: 71) elaborates a number of vital competitive reasons for shortening development times but cites several studies which all show that such a reduction is not a problem that can be solved by a development department approach in isolation. It requires a systems approach, and one needs to create at the very least a close collaboration among development, manufacturing and marketing activities. The greatest savings can be made in the interfaces, in the transfer of technology and information between functions, or in carrying out the different tasks in such a way that the results make the job easier for the other parties. De Meyer produces data showing no evidence of a trade-off between resources and lead time (that is, you can't spend your way to shortened lead time). He also shows the extent of vendor involvement in design and development in Japan and highlights the importance of good vendor relations for fast development times.

He notes techniques for early involvement of other functional groups – job rotation, regular joint review meetings, seminars, joint customer visits, social interactions, and physical proximity of the workplaces. He also notes the use by Toshiba of two types of factory: one for normal mass-production and one, the 'development factory', for developing and bringing through new products into full-scale production (known in the USA as 'ramping up'). He notes that the idea of simultaneous engineering can be found in Abernathy (1971) and that Imai and others (Imai. 1985: 337) identified it as a critical element of success of the Japanese NPD process.

De Meyer concludes by reporting a study of European businesses which shows results that are not very positive. 'European industry does not shine in its development speed, and managers know it. It seems, however, that a healthy reformulation of the manufacturing efforts that can lead to shorter development times is nevertheless under study' (De Meyer, 1992: 81).

We have discussed at some length the 'big three' findings of the Womack and Clark auto industry studies. Clark and Fujimoto note in addition (1991) a number of detailed differences between the Japanese and Western firms. For example, Japanese design engineers go straight to the prototype shop and instruct technicians directly about changes; the US practice is to subject design change orders to a series of formal approvals before allowing them to reach the shopfloor. There are two competing paradigms – 'the prototype as early problem detector' and 'prototype as master model'. The former is Japanese, the latter is European. US engineers frequently expressed dissatisfaction with the level of prototype representativeness. Pilot runs and ramp-ups are relatively short in Japan, partly because they are done in the main production facility, using the normal process workers who learn on the job. Also the number of prototypes made is much smaller – 53 on average in Japan versus 129 in the USA and 109 for volume European producers.

Clark and Fujimoto end by noting the common perspective that design and manufacturing are distinct and radically different activities. They suggest this dichotomy of view is reflected and reinforced in academic research, where there has been focus on the R of R&D – what happens in laboratories – and on production in single, unchanging product/constant technology environments, like that of F. W. Taylor's hod carriers and pig-iron loaders. Where commercial product development is studied it is often of a unique and/or breakthrough kind, for example defence contractors, ethical pharmaceuticals (or computers). Clark and Fujimoto then go on to point out that for most commercial products the design and manufacture is not dichotomous and they point up the similarities – each has physical and informational outputs and each engages in much information generation.

SOME WESTERN ATTEMPTS TO IMPROVE DESIGN FOR MANUFACTURING

Trygg (1990) reports three case studies in Swedish manufacturing – he measured location of departments and other factors and discovered a size-dependent relationship – 'the larger the organization, and hence more complex, the greater the need for effective co-ordination, which consequently leads to greater use of integration mechanisms' (reported in Ettlie, 1992: 111). Ettlie himself has measured integration by using five factors: (1) outside training for the design team in Design for Manufacturing techniques; (2) manufacturing sign-off on design releases; (3) novel structures used in an *ad hoc* way for integration; (4) job rotation practice in engineering functions; and (5) permanent reassignment across functions. He found significantly higher new system utilization when greater integration was accomplished. New structures that included effective teams for integration accounted for a significant amount of the return on investment on these modernization projects, and the minority of firms (15%) that involve workforce participants in the new system design process are the most innovative.

Slusher and Ebert (1992: 131) suggest a contingency approach, arguing that two dimensions, the frequency of problems versus their structuredness, give rise to four cells – incremental design, complex design, creative design and intensive design, each requiring different organizational arrangements for co-ordination. Intensive design (the one in the cell with the highest frequency of problems, each least structured) requires all the features of lean design (organic structure, decentralization, effectiveness goals, integration, direct contact, simultaneous engineering, multidisciplinary teams, team control, liaisons, and team linkages), whereas incremental design (the cell representing the simplest case of design with infrequent problems, each well structured) was best co-ordinated by the traditional mechanistic command-and-control methods.

Adler (1992) pursues the same idea, using a fourfold co-ordination typology of standards, plans and schedules, mutual adjustment, and teams. He then identifies factors which either prevent or encourage firms to learn to use improved DFM co-ordination techniques. Factors *against* include organizational inertia (force of habit), lack of resources, political costs of change, and norms/values/schemas. (Design engineers often see themselves as smarter than manufacturing engineers and therefore behave like prima donnas. They are sometimes recruited from more prestigious schools, and top management often signals in numerous ways that they have a higher status in the organization.) Factors *for* change include business crises ('Not one of the organizations that I have studied has made substantial improvements in their DFM performance without a business crisis' (Adler, 1992: 150)), demands from above, technological pressure, and environmental pressures, for example, from defence procurement agencies.

Change to DFM requires changes in artefacts such as revised reward and promotion criteria, institutionalized DFM assessment methodologies, and 'war rooms' where product–process design teams can meet; changes in values, for example greater trust; and changes in basic assumptions.

Underlining Clark and Fujimoto's contention above, that development and manufacturing are unnecessarily polarized, a study by Susman and Dean (1992) identified many important differences between product and manufacturing engineers: on tolerances, materials versus labour costs, time constraints, state-of-the-art versus possible, materials key to the design or not, jobs vulnerable or secure, slope of career path, salary level, what is rewarded (patents or making schedule), fuzziness or otherwise of performance criteria, creative versus pragmatic orientation. However, contrary to Clark and Fujimoto, Susman and Dean conclude that these differences are an intractable problem and are better tackled by providing a high level of integration between the functions. They suggest that good project outcomes might be obtained if such integration mechanisms as developing group processes, and computerization and codification were employed.

Liker and Fleischer (1992), whose image of factory chimneys we referred to earlier, emphasize the cultural barriers to integration. They too identify incompatible product–process engineering sub-cultures and the low status of manufacturing in the USA, and also add to the list the individualistic values in North American culture.

Their study of two divisions of a US firm trying to eradicate these barriers revealed a list of actions:

- merging product–process reporting lines;
- co-location;
- cross-functional teams;
- combining product and process departments;
- cross-training;
- process designers implementing their own designs as part of an autonomous working group;
- resident product engineer in plant;
- early component supplier sourcing;
- early equipment engineering contract;
- raising the manufacturing pay scale;
- socialization of new recruits into the new culture;
- emphasizing teamwork as a corporate value;
- teamwork training.

They report some measurable success in one division in the newest launch after many of these mechanisms had been put into place. The other division had yet to do a product launch.

Francis and Winstanley (1992) report their study of the management of engineering design in the UK. They note that by the mid-1980s many UK companies had become aware of the uncompetitiveness of their NPD process and in the engineering industry had begun to do something about it by recruiting a much higher percentage of graduate engineers. Over the period 1978 to 1990 the ratio of professional engineers to all staff in the UK engineering industry more than doubled. However, changes to the organization and management of the firms which employed them were much less widespread. In effect, firms were recruiting increased numbers of staff with late twentieth-century skills and putting them into late nineteenth-century organizations. A later survey by Merchant (1991), under Francis's supervision,

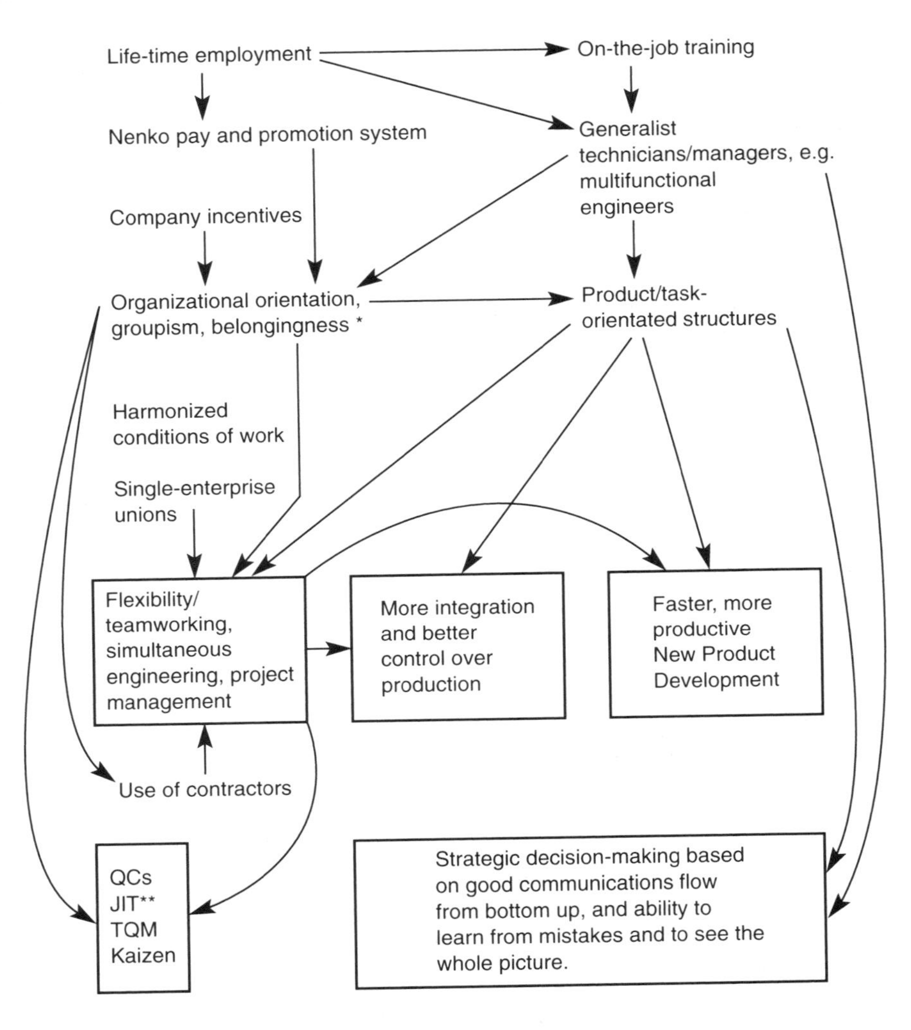

Figure 4.1 Some relationships between aspects of Japanese management practice
* cf. Western occupation orientation or contractual relation with, and independence from, employer. ** By JIT we mean not only Just-in-Time but, following Voss, the emphasis on maintaining flow, flexibility and developing the chain of supply

showed that things had hardly changed between the Francis/Winstanley survey which took place in 1985 and her own survey of 1991.

Moreover, of those rather few firms which had attempted to re-engineer their NPD process, most had experienced some difficulties. In a number of cases firms seem to have attempted simply to bolt on particular elements of Japanese practice to existing organizational and managerial arrangements. Francis and Winstanley suggest that the specific features of Japanese practice that are associated with efficient and effective NPD, prinicipally simultaneous engineering, project management and teamworking, are not optional extras. In Japan they are closely tied into a wider matrix of organizational practices. This is shown in Figure 4.1.

WHY ARE WESTERN FIRMS HAVING SO MUCH DIFFICULTY RE-ENGINEERING THEIR DESIGN FOR MANUFACTURE PROCESSES?

I have argued elsewhere (Francis, 1992), following Best (1990), that UK firms are not just faced with the challenge of making incremental adjustments to their organizational and managerial arrangements, instead they are faced with the need for major institutional change if they are to remain competitive. This is a task that is becoming popularly known as re-engineering (Hammer, 1990).

Best's argument periodizes industrial history into three phases. The first is that following the industrial revolution during which craftwork predominated. The second dates from the early part of the twentieth century and is characterized by what Best terms 'the American system of manufacture' – based on the development of the standardization of parts and mass-production, and the rise of modern management, control passing from the hands of the craftworker and into the hands of the production engineers. The third phase postdates the Second World War and is characterized by those elements of Japanese management practice we have outlined above, and also by the way in which work gets done in industrial districts such as Emilia-Romagna in Italy. Best terms this third phase that of 'the New Competition' and identifies four of its dimensions. Firms are organized as collective entrepreneurs rather than on a command-and-control basis. The production chain is organized neither in a hierarchic/bureaucratic fashion nor by market relations between autonomous independent firms. Instead there is consultation/co-operation amongst mutually interdependent firms. Industrial sectors are characterized by a substantial amount of 'extra-firm infrastructure' (Piore and Sabel, 1984), and fourthly there is a carefully crafted industrial policy.

A crucial part of this argument is that nations find it extraordinarily difficult to make the transition from one epoch to the next. Specifically Best suggests that the UK, which pioneered the industrial revolution, failed to adopt successfully the American system of manufacture and the US has failed to adapt to 'the New Competition'. This belief that firms, and therefore aggregations of firms in the form of a nation's industry, get locked into the organizational forms and managerial practices that were state-of-the-art at the time of their formation has a long and respectable intellectual history, from Stinchcombe (1959), through the population ecologists (see, for example, Hannan and Freeman, 1974, 1989), to Lazonick (1990).

If one takes this view, several important things follow for the competitiveness of Western firms in the management of their design–manufacturing interface. Firstly, the task facing management, and technical staff, in the individual firm is daunting. They are not just facing an organization or culture change programme amenable to the techniques of Organization Development (OD) or other change consultants. The implication of this analysis is that almost everything has to change. Occupational categories which were invented for nineteenth-century craftsmen and draughtsmen (the gender specificity is intentional!) need to be replaced by a looser, more flexible, way of labelling the work people do. Hierarchical organizational forms, designed to command and control, need to be replaced with co-ordination arrangements that allow more initiative and autonomy to be exercised at lower levels of the organization. New ways of preventing free riding, providing incentives, and co-ordinating complex tasks have to be invented and institutionalized. A new architecture (Kay, 1993) of relationships both within and between firms has to be designed, built and become taken-for-granted. And all this is bound to take a very long time – not just because of the size of the task itself and the time people take to change their mind-set, but because the 'New Competition' requires long-term relationships of commitment and reciprocity to be built up, otherwise game playing, free riding and opportunism (Williamson, 1985; Kay, 1993), previously prevented by command-and-control organizational forms, will undermine the new arrangements.

Secondly, firms cannot do this by themselves. We have here what seems to be a clear case of market failure – on two counts. The one is that the provision of the infrastructural arrangements and the development of an appropriate industrial policy is beyond the capacity of the individual firm to achieve. The other is that the move to the inter-firm architecture of the 'New Competition' may not be possible without outside help.

Taking these in turn, even if firms in an industry were used to co-operating with each other they may find it difficult to set up on a joint basis the kind of infrastructural arrangements Best identifies as typical of the 'New Competition' (industry training programmes, joint marketing arrangements, regulatory commissions, trade associations). Some kind of extra-firm organization is needed to develop and maintain these institutions. If one goes further, and believes, as does Best, that the new competition requires an industrial policy which shapes and uses the market, which (as in Japan) focuses on the supply side by setting up consultative buyer–vendor relations, inter-firm associations, and extra-firm agencies that facilitate continuous improvement in production, and which targets strategic sectors to maximize industrial growth, it is clear that firms cannot pull themselves by their bootstraps from one industrial epoch to the next.

If, in addition, it is the case that firms have a history of acting in an individualistic fashion, used to arm's-length zero-sum game relations with their suppliers and competitors, then it is even more unlikely that without outside intervention they will be able to set up the above mechanisms and policies.

Furthermore, it will be very hard for any one firm to be first mover away from this kind of atomistic architecture characteristic of the American system of manufacture to set up consultative–co-operative relations with mutually interdependent firms typical of the 'New Competition'. This is a classic 'market failure' scenario requiring non-market intervention to set up the new architecture. In industries dominated by a very small number of very large firms it may be possible for one or more of those firms to act decisively in a way that reshapes that industry. Otherwise the intervention of an outside agency may be necessary.

CONCLUSIONS

Three conclusions therefore follow. The first is that Design for Manufacture is a crucial element in a firm's competitive strategy. The research reported here shows how it is done well in Japanese firms, gives some indication of the comparative success it provides, and paints a dismal picture about the extent to which this best practice is not followed in the West.

The second is that many Western firms face the need to make a step-change, not an incremental change, in the way they organize and manage their New Product Development process. But this step-change is not a matter of simply bolting certain aspects of good Japanese practice on to existing organizational arrangements, the change needs to be more radical and thoroughgoing. The broad shape of that change seems reasonably well established.

The third is that it is unlikely that individual firms will be able to make the necessary changes alone. They are faced with a classic case of market failure. New institutions need to be created beyond the level of the individual firm. Moreover, individual firms need to move closer to each other but in many cases there may not be enough incentive for any one firm to make the first move. What agencies might act to overcome this market failure? In the USA it seems that the Clinton administration is set to address this issue seriously and purposefully. In the UK likely candidates, in theory at least, are the Department of Trade and Industry, Scottish Enterprise, the TECs and LECs, the CBI through its Manufacturing Council, and industry associations. Can they rise to the challenge?

If they cannot, there are two hopes one might express. The one is that the high level of awareness of the fall in competitiveness of the manufacturing sector, at least in the UK and the USA, is bound to lead to some action, and such action as is taken will follow a route that is sensitive to local social, political and economic considerations. Slavishly following the Japanese pattern may not prove to be the best strategy. Secondly there is the possibility that Japanese companies may not be able to keep up their present arrangements. We are already seeing a marginal retreat from the strategy of product proliferation that has been so successful until recently. The downturn in the Japanese economy has reduced the demand for such a high level of product variety, and the cost of producing that variety is now viewed as too high during a recessionary period. Furthermore, the high level of commitment to the team and to the enterprise, characteristic of the Japanese worker, may have been a response, to some extent, not just to a rational way of running a 'New Competition' enterprise, but to being a man of a certain age in a society with a certain history. Whether the younger generation of Japanese, of both sexes, born and bred in the relative affluence of the post-1960s boom years in Japan will share the same values as the 1930s- and 1940s-born *salarimen* is an empirical question. There is already a Japanese word for the younger generation, the *shinjinrui* – the new breed. But if I were the chief executive of a UK engineering company I would not take too much comfort from this knowledge.

REFERENCES

Abernathy, W. J. (1971) Some issues concerning the effectiveness of parallel strategies in R&D projects, *IEEE Transactions on Engineering Management*, EM- Vol. 18 no. 3.

Adler, P. S. (1992) Managing DFM: Learning to Coordinate Product and Process Design in G. Susman (ed.) *Integrating Design and Manufacture for Competitive Advantage*, Oxford University Press, New York.

Barkan, P. (1992) Productivity in the process of product development – an engineering perspective, in G. I. Susman (ed.) *Integrating Design and Manufacturing for Competitive Advantage*, Oxford University Press, New York.

Best, M. H. (1990) *The New Competition: Institutions of Industrial Restructuring*, Polity Press, Cambridge.

Clark, K. B. and Fujimoto, T. (1991) *Product Development Performance: Strategy, Organization, and Management in the World Auto Industry*, Harvard Business School Press, Boston, Mass.

Clark, K. B., Chew, W. B. and Fujimoto, T. (1992) Manufacturing for design: beyond the production/R&D dichotomy, in G. I. Susman (ed.) *Integrating Design and Manufacturing for Competitive Advantage*, Oxford University Press, New York.

De Meyer, A. (1992) The Development/Manufacturing Interface: Empirical Analysis of the 1990 European Manufacturing Futures Survey in G. Susman (ed.) *Integrating Design and Manufacture for Competitive Advantage*, Oxford University Press, New York.

Ettlie, J. E. (1990) Methods that work for integrating design and manufacturing, in J. E. Ettlie and H. W. Stoll (eds.) *Managing the Design–Manufacturing Process*, McGraw-Hill Book Co., New York, pp. 53–77.

Ettlie, J. E. (1993) Concept Development Effort in Manufacturing in G. Susman (ed.) *Integrating Design and Manufacture for Competitive Advantage*, Oxford University Press, New York.

Francis, A. (1992) The process of national industrial regeneration and competitiveness, *Strategic Management Journal*, Special Issue, Winter 1992.

Francis, A., Snell, M., Willman, P. and Winch, G. (1982) The impact of information technology at work: the case of CAD/CAM and MIS in engineering plants, in L. Bannon *et al.* (eds.) *Information Technology: Impact on the Way of Life*, Tycooly Press, Dublin.

Francis, A. and Winstanley, D. (1992) The organization and management of engineering design in the UK, in G. Susman (ed.) *Integrating Design and Manufacture for Competitive Advantage*, Oxford University Press, New York.

Hammer, M. (1990) Re-engineering work: don't automate, obliterate, *Harvard Business Review* Vol. 68, no. 4, July–August, pp. 104–12.
Hannan, M. T. and Freeman, J. (1974) Environment and the structure of organizations. Paper presented at the annual meetings of the American Sociological Association, Montreal, Canada.
Hannan, M. T. and Freeman, J. (1989) *Organizational Ecology*, Harvard University Press, Cambridge, Mass.
Imai, K., Nonaka, I. and Takeuchi, H. (1985) Managing the New Product Development Process: How Japanese Factories Learn and Unlearn in K. B. Clark, R. H. Hayes and C. Lorenz (eds.) *The Uneasy Alliance: Managing the Productivity – Technology Dilemma*, Harvard Business School Press, Boston.
Kay, J. (1993) *Foundations of Corporate Success: How Business Strategies Add Value*, Oxford University Press, New York.
Knight, D. (ed.) (1977) *Matrix Management*, Gower Press, Aldershot.
Lazonick, W. (1990) *Competitive Advantage on the Shop Floor*, Harvard University Press, Mass.
Liker, J. K. and Fleischer, M. (1992) Organizational context barriers to DFM, in G. Susman (ed.) *Integrating Design and Manufacture for Competitive Advantage*, Oxford University Press, New York.
Merchant, L. (1991) Organizational arrangement for New Product Design. MBA dissertation, Imperial College.
Piore, M. and Sabel, C. (1984) *The Second Industrial Divide*, Basic Books, New York.
Rosenthal, S. R. and Tatikonda, M. V. (1992) Competitive advantage through design tools and practices, in G. Susman (ed.) *Integrating Design and Manufacture for Competitive Advantage*, Oxford University Press, New York.
Sanderson, S. S. (1992) Design for manufacturing in an environment of continuous change, in G. Susman (ed.) *Integrating Design and Manufacture for Competitive Advantage*, Oxford University Press, New York.
Slusher, E. A. and Ebert, R. J. (1992) Prototypes for managing engineering design processes, in G. Susman (ed.) *Integrating Design and Manufacture for Competitive Advantage*, Oxford University Press, New York.
Stinchcombe, A. (1959) Bureaucratic and craft administration of production: a comparative study, *Administrative Science Quarterly*, Vol 4, pp. 168–87.
Susman, G. I. and Dean, Jr, J. W. (1989) Strategic use of computer-integrated manufacturing in the emerging competitive environment, *Computer-Integrated Manufacturing*, Vol. 2, no. 1, August, pp. 133–8.
Susman, G. I. and Dean, Jr, J. W. (1992) Development of a model for predicting design for manufacturability effectivness, in G. Susman (ed.) *Integrating Design and Manufacture for Competitive Advantage*, Oxford University Press, New York.
Trygg, L. (1990) The use of integration mechanisms in the design to production transfer. Working paper no. 1990:31, presented at the International Conference on Advances in Production Management Systems, 20–22 August, Espoo, Finland.
Williamson, O. E. (1985) *The Economic Institutions of Capitalism*, Free Press, New York.
Womack, J. P., Jones, D. T. and Roos, D. (1990) *The Machine that Changed the World*, Rawson Associates, New York.

5

FLEXIBLE MANUFACTURING SYSTEMS

Harry Boer

INTRODUCTION

Industries worldwide have been confronted with a number of intertwined changes in their environments. Many industrial markets have turned into virtually worldwide 'battlefields' in which customers are demanding ever-wider ranges of relatively low-cost, reliable and high-quality products, and ever-shorter and reliable delivery times. At the same time, a range of Advanced Manufacturing Technologies and innovative forms of organization have become available which are believed to contribute considerably to their adopters' capability to meet these present demands for cost-effectiveness, quality and flexibility simultaneously. Many companies' futures depend on their ability to adopt and implement these new technologies and novel forms of organization successfully.

In Chapter 7 several forms of Advanced Manufacturing Technology are presented. One of the most promising of these forms is the so-called Flexible Manufacturing Systems (FMS). The purpose of this chapter is to discuss in more detail the organizational aspects of the implementation and operation of FMS. It is shown that the decision to adopt FMS entails a manufacturing innovation, a term coined by Braun (1981) and used to denote that the innovation comprises much more than installing just another piece of equipment and putting it into operation by pressing the button. Rather, the implementation and operation of FMS require a wide range of highly interdependent organizational and technological adaptations which are needed in order to ensure that the FMS benefits pursued are fully realized.

The chapter starts with a description of the technological and industrial roots of FMS technology. Next, the typical characteristics and benefits of FMS are discussed. Then, the obtainability of these benefits is addressed and economic, technical and market-related problems preventing or delaying FMS benefits are discussed. The core of the chapter concerns *the organizational conditions* enhancing the success of FMS implementation and

Case	Size*	Products and batch sizes	Market demands	Type of FMS	Year**	FMS capabilities
A	1,900	Coal-mining equipment	Product quality and delivery time	6 CNC MCs, rail-guided AGV***	1984	1,000 parts per year (mainly major castings), 30 variants; replaces 16 conventional machine tools and 2 CNC MCs
B	200	Hydraulic equipment	Price, delivery time and product modification	2 cells, linked by conveyor; cell 1; 3 CNC MCs, 1 measuring machine, 1 washing station. 1 robot; cell 2; pressing/drilling equipment, 1 robot	1984	100,000 sets of parts per year for 2 main types of pump; conventional manufacture would require a line of 9 machines
C	1,050	Diesel engines	Price, product quality and delivery time	3 CNC MCs, rail-guided AGV	1985	35,000 parts per year; 3 main types; water and exhaust manifolds, and thermostat housings, 200 variants; replaces 12 conventional drilling, boring and milling machines
D	1,050	Diesel engines	Price, product quality and delivery time	9 cells, linked by conveyor system; cell 1: 4 CNC MCs, 1 AGV, 1 robot; cell 2: 6 CNC MCs, 1 saw, 1 wash, 1 robot; cell 3: 5 assembly stations, 1 robot; cell 4/5; 1 robot, 1 CNC MC, 3 borers, 1 wash, 1 gauge; cell 6: 2 robots, 6 assembly stations; cells 7, 8, 9: conventional equipment	1987	14,000 rods per year; 4 types; replaces a line of 25 conventional drilling, boring, milling and other machines
E	700	Printing and finishing systems for wallpaper, textile and floor-covering industries	Product innovation and quality	2 CNC MCs, rail-guided AGV	1985	42,000 parts per year, 315 variants, comprising all cubic parts of final products; replaces 4 CNC MCs and 1 CNC milling machine
F	400	Truck engines	Delivery time and product modification	6 CNC MCs, rail-guided AGV	1985	30,000 cylinder blocks and heads per year; 4 main types; replaces part of 4 conventional transfer lines, containing 100 machining stations in total
G	400	Truck axles	Delivery time and product modification	2 CNC MCs, overhead robot conveyor	1985	18,000 rear-axle bodies per year, 6 types, 64 variants; replaces 20 conventional machine tools
H	400	Compressors and air conditioners for industrial refrigerating systems	Product modification, price and delivery time	2 CNC MCs each having a pallet carousel	1983	36,000 parts per year, mainly major parts for a family of compressors; replaces 10 conventional drilling, boring and milling machines

Figure 5.1 Summary of company characteristics and FMS features.
Notes: Number of employees at the start of the case study; ** Year of implementation of the FMS; *** CNC = Computer Numerical Control; MC = machining centre; AGV = Automated Guided Vehicle

operation and it is concluded that the benefits of FMS will not be achieved, or only partially, if the adopter is not aware of the necessity to implement these conditions, or not prepared to implement them.

Throughout the chapter, examples are used which are based on longitudinal case studies into the adoption, implementation and operation of FMS by four British, one Belgian and three Dutch companies. Earlier reports by the present author (see, for example, Boer, 1991) have described these case studies in much more detail. In Figure 5.1 the main characteristics of the companies involved and of the FMS adopted by them are summarized. Also, data are used based on other FMS studies reported in the literature.

FLEXIBLE MANUFACTURING SYSTEMS

Technological roots and definition

This section will introduce the main characteristics of Flexible Manufacturing Systems. The description is based on Groover and Zimmers (1984), Macbeth (1985), Bessant (1989) and Hill (1989). Flexible Manufacturing Systems (FMS) are widely regarded as a major step towards 'the factory of the future'. They combine several items of hardware, software applications and controls based on a number of technological developments in flexible manufacturing, in particular:

1. Group Technology, in the 1950s;
2. Numerical Control, in the late 1940s, followed, in the 1960s and 1970s by Computer Numerical Control and Direct Numerical Control;
3. machining centres, which first appeared in the late 1950s;
4. transport, handling and storage systems, including industrial robots, Automated Storage and Retrieval Systems, and Automated Guided Vehicles; all started in the 1960s but really took off in the 1970s.

Group Technology (GT) is a manufacturing philosophy in which certain parts are identified and grouped together to take advantage of their similarities in manufacturing. The benefits of GT include standardization of tooling and set-ups, reduced material handling, lead times and Work in Progress, and improved delivery reliability. The most important characteristic of GT, in this context, is that it provides the basis for organizing facilities into cells containing all the machines needed to produce a family of parts. FMS are an example of such cells.

Usually, machine tools in FMS are Numerically Controlled. Numerical Control (NC) technology, which evolved out of a program in the US Air Force in the late 1940s, refers to the automated operation of machine tools according to a detailed set of coded instructions, in the form of an NC part program. These instructions are based on numerical data stored on paper or magnetic tape, punched cards, or computer storage. The key difference between NC technology and conventional machine tools is that of reprogrammability. In NC, design changes and modifications can be carried through simply by changing the instructions. Thus, the human component in controlling the machine tool, and in translating design information into operations, is replaced by a control device. Compared with conventional equipment, NC machines offer increased accuracy, consistency and flexibility, the trade-off being the increased investment associated with NC equipment, and changes required in skill requirements.

NC technology did not diffuse widely until the late 1960s when low-cost reprogrammable controls became available. In Computer Numerical Control (CNC), the hard-wired control unit of the NC system is replaced with a stored program, using a microprocessor. The advantages of CNC, as compared with NC, include even greater flexibility and reliability, mainly through the use of a memory storage instead of paper-tape input. Initially, the application of CNC was restricted to automating machining operations such as milling, boring and turning, but soon other NC-controlled operations became possible, such as tool change and also parts storage, handling and transport, using, for example, robots or rail- or wire-guided vehicles. One step further are machining centres in which CNC operations previously provided by different machines are combined into one machine comprising a tool magazine to enable automated tool change.

The next step, in the early 1970s, based again on advances in computer technology, involved the development of Direct Numerical Control (DNC) in which a number of CNC tools could be grouped into a manufacturing cell under overall control of a host computer controlling scheduling and routeing of parts, monitoring status, feeding new programs, and so on.

From this the step towards so-called Flexible Manufacturing Systems (FMS) was a relatively small one. An FMS is a group of NC machines or other workstations which are interconnected by an automated material handling and storage system. The system operates under central computer control using DNC and is capable of processing a variety of different types of parts simultaneously under NC control at various workstations. Most FMS consist of a mixture of standard and one-off elements.

Industrial roots and application

The companies referred to in this chapter are typical examples of the types of situation for which FMS technology was developed in the first place (see Figure 5.1). Companies B, D and F, for example, are high-volume producers which operated a line of dedicated equipment for parts production. In companies A, C, E and H the FMS replaced general-purpose machinery used for low- or medium-volume batch production of parts. Finally, company G is a medium-volume manufacturer which used a line of relatively flexible equipment prior to installing an FMS.

Designed to fill the gap between high-production transfer lines and low-volume stand-alone NC machines, FMS are supposed to provide adopters with the opportunity to produce a family of parts both flexibly and cost-effectively (see Figure 5.2).

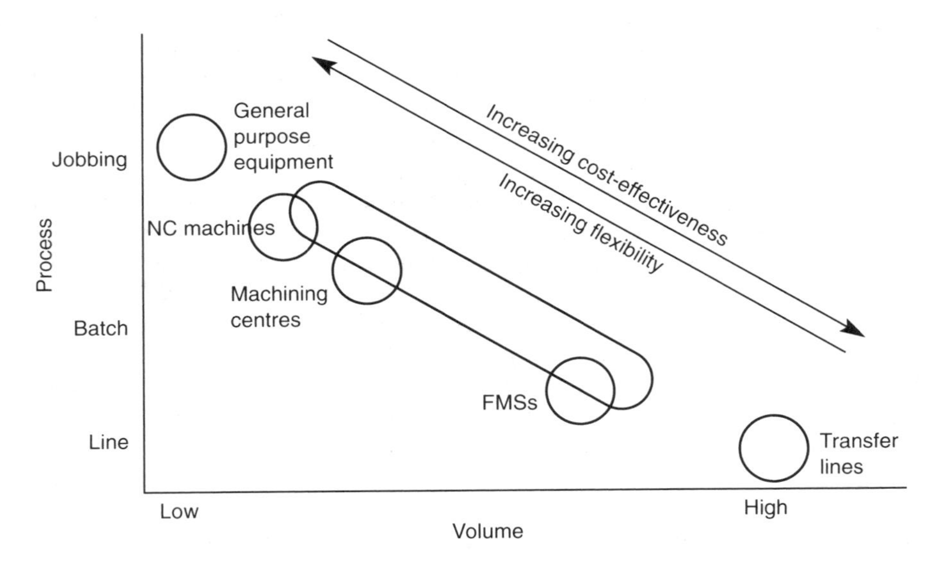

Figure 5.2 The position of FMS and the technologies they are based on, among the classic choices of process. (From Hill, 1989)

Western industries have traditionally regarded this combination of flexibility and cost-effectiveness as incompatible with their manufacturing capabilities. This has resulted in a trade-off between the high-volume manufacture of standard products, using special-purpose equipment, and low-volume manufacture of batches of a wide variety of parts, or even one-offs in job shop manufacture, using general-purpose machinery. Both options have their drawbacks.

A dedicated transfer line is expensive to install and sensitive to breakdowns. Mass-producers tend to perform well on cost-effectiveness and, delivering from stock, delivery reliability is high, and lead times and delivery times are low. However, their product flexibility is limited. If product design is changed, at least some parts of the present equipment are likely to become obsolete. This type of process has increasingly come under pressure, as product lives were shortening so that the frequency with which new models were introduced increased. At the same time the number of variants demanded by customers increased, too. Consequently, the pressure on these types of manufacturer to improve their flexibility increased.

Low-volume, high-variety manufacturers, in contrast, are an even more important sector in Western industries. About 75% of European manufacturers produce batches of an average size of thirty to forty units. This type of firm is highly flexible through the use of general-purpose machine tools organized on a functional layout. At the same time, however, they are characterized by large Work in Progress stocks between operations because of high set-up costs. Levels of actual manufacturing time are extremely low (utilization degrees of 5–10% are not uncommon) and so is the number of stock turnovers per year. These systems cannot react rapidly to the increasing pressure on cost-effectiveness, delivery time and delivery reliability, while maintaining adequate levels of product flexibility. Consequently, this type of industry has looked for alternative ways of producing both flexibly and cost-effectively.

The Japanese have shown the way by developing so-called 'lean' manufacturing organizations: market-oriented combinations of novel ways of organizing people and production processes, superior systems of quality control, production planning and maintenance, and advanced forms of flexible automation such as robots and FMS. The lesson that can be learned from the Japanese is twofold. Firstly, flexible automation is by no means the panacea for obtaining, sustaining or increasing success in the marketplace. Secondly, if flexible automation is incorporated in the production system, it is hardly ever the mere implementation of technology but rather the combination of automation, organizing people and processes, and implementing adequate operations management systems, that really improves performance in the marketplace.

Typical characteristics of FMS technology

In Figure 5.3 the characteristics that distinguish FMS from conventional equipment are summarized. The main characteristics are the integration, mechanization and reprogrammable automation of operations (parts machining, material handling and tool change), technical flexibility, complexity, regulation and expensiveness (Boer, Hill and Krabbendam, 1990). In Figure 5.4 the relationships between these factors are shown.

The way these FMS characteristics operate can be explained as follows. Generally, the use of an FMS implies the integration of previously independent operations. Human activities are mechanized or even automated as are

Characteristics	Definition
Integration	The extent to which the system is able to perform different types of operations, to change tools, to transfer and load workpieces, and to download part programs and production schedules from a central storage
Technical flexibility	The ability to quickly change mix, routeing and sequence of operations within the parts envelope
Mechanization and reprogrammable automation	The degree to which operations such as workpiece transfer, loading/unloading and fixturing, tool change, machine tool control, cutting tool control and quality control, are performed by the system, without human intervention
Complexity	The number of inter-related elements comprised in the system, such as workstations, material-handling system, control system, and other elements
Regulation	The extent to which the system regulates the work of operators, process planners, production planners, maintenance engineers and other personnel
Expensiveness	The costs incurred in the investment in, and the operation, maintenance and operational management of the system

Figure 5.3 Characteristics of FMS technology

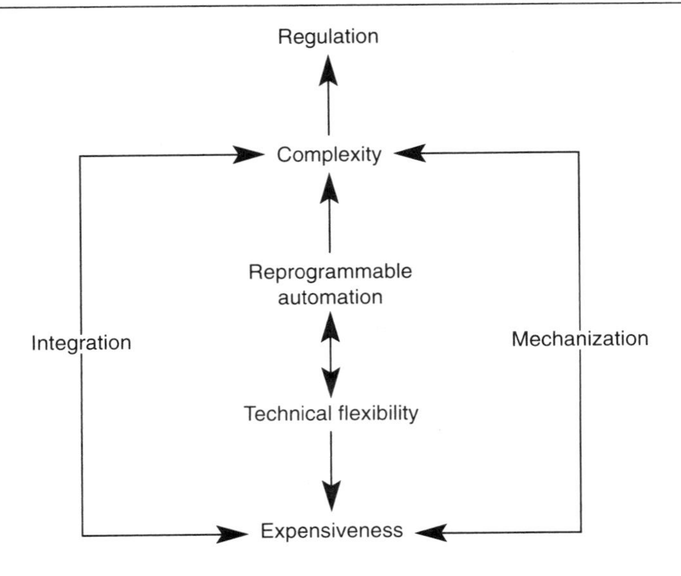

Figure 5.4 Relationships between the FMS characteristics

most mechanical operations. All of this results in a reduction in the amount of equipment and the number of operators needed to produce a certain volume.

The technical flexibility of FMS concerns the ability to quickly change mix, routeing and sequence of operations within the parts envelope. Based on this feature and that of reprogrammability, FMS may contribute to improving its adopter's operational and strategic flexibility in four ways. The degree to which set-up flexibility increases depends on whether the FMS replaces dedicated or more flexible machinery. Set-up flexibility strongly affects the obtainability of operational benefits such as reduced batch sizes, lead times and Work in Progress (see next sub-section). Furthermore, FMS may contribute to improving its adopter's modification flexibility: a more strategic advantage. Provided that no additional tools or fixtures are needed, modification is just a matter of changing part programs that have already been written. Also, product flexibility is affected positively, as long as new designs keep within the capabilities of the system. Sizes, tolerances, materials or other features falling outside the system envelope cannot be produced, unless the system is modified which, however, may come very costly. Finally, the contribution of FMS to increasing volume flexibility is limited, which may make the system rather sensitive to changing market demand, as will be discussed at more length in a subsequent section of this chapter.

FMS are very complex systems. They comprise a large number of different elements based on a variety of technologies that operate in a highly interactive way. Due to this and the level of automation, FMS is a highly regulating technology. Requiring uniform behaviour as regards data input, parts fixturing, change of worn tools and predictable reactions to disturbances and breakdowns, FMS increases the number of rules and procedures to regulate the employees' behaviour. The cost-effectiveness of FMS gives a rather mixed picture. Due to the mechanization and automation of operations, manning levels tend to decrease, thus reducing the costs of operating the system. On the other hand, process planning costs tend to increase. Furthermore, using the much increased set-up flexibility, FMS adopters are able to reduce set-up costs. Usually, however, investment and maintenance costs of fixturing devices increase, and so do other maintenance costs as a wider range of more expensive spare parts needs to be held in stock than before. So, the feasibility of FMS depends a great deal on the balance of costs and benefits achieved.

FMS benefits

Many authors have emphasized the strategic advantages of FMS. In other publications it has been suggested that FMS adopters need not necessarily pursue market benefits, but may 'just' be aiming at reducing operation times and costs and improving production management. Our research (Boer, 1991) seems to explain the latter observation (see Figure 5.5). The benefits identified in the literature (see, for example, Charles Stark Draper Laboratory, 1984; Hartley, 1984; Groover and Zimmers, 1984; Bessant and Haywood, 1988) can be classified as follows (Krabbendam, 1988):

1. *Improved market performance:* a more adequate and rapid response to market demands for increasing product diversity and innovation; lower sales prices; shorter delivery times and higher delivery reliability; and improved quality.
2. *Reduced costs/times of operations:* reduced labour, overhead costs and floor space; shorter processing, set-up and manufacturing lead times; reduced batch sizes and Work in Progress; increased machine utilization.
3. *Improved operations management:* linking of production control and manufacture; increased scheduling flexibility; Just-in-Time manufacture; fewer human errors; and improved and consistent quality and productivity.

Among the major *spin-offs* reported in the literature are: standardization of product designs, tools and fixtures; increased knowledge of Advanced Manufacturing Technology; improved logistics and quality management; a step towards (Computer)-Integrated Manufacture; and improved company image.

In spite of these benefits, FMS are still a rare phenomenon in companies. The first FMS was developed in the early 1960s in the UK. By 1981 the FMS population had grown to slightly over 100. In 1985 it was estimated that well over 1,000 FMS were in use worldwide. (See, for example, Bessant (1989). These figures are just meant to give an indication. Estimates vary widely,

With the exception of reduced delivery time, which was pursued by most companies, operational advantages were emphasized, rather than market advantages. It was only for company A that market advantages, i.e. improved quality and reduced delivery time, were the prime motive for investing in FMS. In case C the company felt it could best achieve its main objective, namely cost reduction through reduced Work in Progress and Just-in-Time delivery to assembly, by using an FMS. For companies B, F and G the prime motive was the launching of a new range of products which the companies felt could be produced more cost- effectively and flexibly using an FMS. In company F, replacement of facilities was also necessary for technical reasons. This also applied to companies C, E and H which had to find alternatives for technically obsolescent machinery.

Most companies' first concern was to justify a technology they considered most suitable for solving their initial problem and the operational advantages pursued actually served as criteria to financially evaluate alternative technologies, rather than being goals in themselves. The most important benefits mentioned across the sample were reduced direct labour, increased machine utilization, reduced set-up times, smaller batches, and reduced Work in Progress.

Research by Bessant (1989; see Figure 5.6) among fifty FMS adopters in the UK confirms this observation which can be explained as follows. In all cases but company B the project was dominated by process engineers with the other functions playing a secondary role. Due to the process engineers' position within their respective companies, operational benefits were emphasized as a basis of justification, rather than competitive advantages which are much more difficult to assess and fall outside the scope of the process engineers. As a result, however, most of the organizational adaptations were aimed at achieving the operational goals pursued, and minor attention, if any, was paid to the market advantages of FMS.

Figure 5.5 FMS in practice: benefits pursued

and depend on the definition of FMS used.) Until recently it was expected that the FMS market would grow quite rapidly. However, growth is lagging behind expectations due partly to unfavourable economic circumstances in many industrialized countries. Another reason is the problematic nature of FMS which has given the technology a poor press. FMS are complex systems and many obstacles must be overcome in order that the benefits of this technology can be achieved.

ACHIEVEMENT OF BENEFITS

Publications addressing success and failure of FMS implementations leave some doubt as to how the promises of FMS are realized in practice. Early adopters were reported to have many difficulties with achieving the market advantages arising from increased flexibility, such as numerous product options and frequent modification. Jaikumar (1986), for example, reported that most FMS installed in the USA showed an astonishing lack of flexibility, and were used for high-volume production. According to Tombak and De Meyer (1988), European and US firms did not adopt FMS to rapidly change their product designs, but to standardize their product lines, using FMS to cover the variability of inputs. More recent reports, however, have shown that many of the operational benefits of FMS, as well as market advantages such as shorter lead times and improved product quality, may indeed be obtained (for example, Bessant and Haywood, 1988; Boer, 1991; also see Figure 5.6) provided, though, that certain barriers are reduced or, even better, prevented.

These barriers, which may delay or even prevent the full benefits of FMS, seem to fall into five broad categories. We shall discuss each of the categories and start by saying a few words about different problems related to the *financial appraisal* of FMS. Next we shall refer to some *technical* problems that may occur. Yet another category concerns the impact of *changes in market demand* taking place during the implementation process. In a separate section we shall discuss the core issue of the present chapter which concerns a range of *organizational conditions* needed for the successful operation of FMS. Many of the FMS benefits can be obtained only if the adopter's organization is sufficiently prepared and adapted to the new technology. Finally, we shall pay some attention to *the problematic nature of the FMS implementation process,* which constitutes a manufacturing innovation process for many adopters. In-depth, processual research of this area has been rare and the reports aimed at filling the gap in knowledge about this type of innovation are relatively few and very recent (for example, Voss, 1988; Gerwin, 1988; Boer *et al.*, 1990; Boer, 1991).

Financial appraisal

The first group of barriers is related to the financial appraisal of FMS and the determination of post-installation operating costs. FMS are very expensive (see Figure 5.6) and involve both higher costs and risks as compared with current technology. Some of the benefits and implementation costs are readily quantifiable in advance; other benefits are qualitative in nature and more difficult to estimate (for example, Kaplan, 1986). Some authors have argued that analytical investment selection tools such as return on investment and discounted cash-flow techniques are of limited use (Blumberg and Gerwin, 1984). Other authors, however, have argued that it is not the methods that are to blame, but the way in which the methods are used by management (Kaplan, 1986; Finnie, 1988) if they are used at all. For example, Primrose and Leonard (1985) devised an interactive computer program for quantifying the intangibles of new technologies at the appraisal stage, based on

	Average	(Range)
Cost of the system (£m)	2.4	(0.8–10.0+)
Number of employees	2,100	(50–16,000)
Batch size	120	(1–2,000)
Part family	240	(2–4,000)
Lead time	–74%	(40–90)
Work in Progress	–68%	(25–90)
Stock turnover	+350%	(50–500)
Machine utilization	+63%	(15–100)

Figure 5.6 Characteristics of fifty UK FMS adopters and benefits achieved by them. (Based on Bessant, 1989)

traditional discounted cash-flow techniques. But a survey reported by Lunt and Barclay (1987) showed that the production directors in over 60% of the companies that had purchased the package were not even aware of the package or who in their company was making use of it! According to Kaplan (1986), managers should avoid claims that investment in Advanced Manufacturing Technology must be made on faith alone. The challenge for managers is to improve their ability to estimate the costs and benefits of new technology, rather than to take the easy way out and discard the necessary discipline of financial analysis.

A major factor underlying these problems is that conventional costing systems are incapable of accurately assessing qualitative factors (Hill, 1985). Consequently, data needed to assess expenditure and income are either inaccurate in many companies or even unavailable. For example, many companies do not have insight into the true nature of manufacturing innovation. Therefore they cannot really know, in advance, the expenditure involved in, for example, reorganization, employee training, additional equipment to be purchased, the use of consultants, and other costs incurred in managing and implementing the manufacturing innovation. All these factors, however, need to be assessed prior to the investment decision as they play an important role in whether or not the investment is justified.

Problems in the pre-adoption appraisal of the technology have resulted in lower rates of diffusion than expected. And many companies that adopted FMS had difficulties with the post-installation assessment of the contribution of the system to the company's performance. However, potential adopters of FMS could learn from other companies' experience. The checklist of technological and organizational conditions needed for the effective operation of FMS discussed in a subsequent section and summarized in Figure 5.10 could be of help here, together with other checklists presented in the literature (see, for example, Groover and Zimmers, 1984; Hartley, 1984).

A very serious problem affecting both the diffusion and the implementation of FMS concerns the short-term view of many managements, imposed upon them by shareholders and financial institutions who prefer short-term return on investment rather than long-term growth (Bessant and Haywood, 1986). This attitude has caused many adopters of new technology to emphasize readily quantifiable benefits in formal financial appraisal. Consequently, many adopters' efforts are primarily aimed at obtaining operational benefits – in particular, reduction of operational times and costs and other benefits that can be quantified – rather than using the new technology to improve the company's performance in the marketplace (Jaikumar, 1986; Haywood and Bessant, 1987; Tombak and De Meyer, 1988). This attitude has a serious effect on the approach of FMS adopters towards the implementation process and the consequences of that approach. In a subsequent section we shall illustrate this issue.

Company A was operating part of its FMS in CNC mode because of faults in the construction of one of the machining centres. This machine (the bottleneck in the parts flow through the system) needed full-time attention, so anticipated reduction of direct labour had been achieved only partially.

Company H experienced all the problems one can think of in relation to a first-generation manufacturing system. Eventually, the system was replaced by a completely new, third-generation system, obtained from the same supplier. To date, however, it is still not clear whether or not the technical problems are finally solved.

Figure 5.7 FMS in practice: technical problems

Technical problems

Many present users of FMS were so-called innovators or early adopters, who experienced all the teething troubles related to the specification and operation of an immature technology. FMS generally consist of a combination of proven CNC machining centres and entirely new, often customized, components. Many adopters had difficulties with these new components, which manifested themselves in, for example, (system) software bugs; the speed of the transport system falling short of specification; communication problems between the machining centres, their tool magazine and the transport system; and problems with automated fixturing devices (Boer, 1991). Figure 5.7 illustrates some of the technological problems encountered by the companies involved in our research. Also, somewhat surprisingly, problems with the more mature elements of FMS were encountered more frequently than expected.

Many problems were related to the immaturity of FMS, such as low levels of standardization, lack of software protocols and insufficient compatibility of different elements, and have been reported to have arisen well into the operating life of many systems (Bessant and Haywood, 1986). Furthermore, during the early days of FMS only a few suppliers had sufficient expertise in both computer applications and mechanical engineering. Also, many of the early FMS adopters lacked the knowledge to specify the most suitable system for their situation and to operate and maintain the system after installation (Gerwin, 1988; Boer, 1991).

Present adopters should experience far fewer problems of this nature. Experience with the design and operation of FMS has grown. Furthermore, standards such as General Motors' Manufacturing Automation Protocol (MAP) and developments in the area of Open System Architecture have made the design and integration of FMS less complicated. Yet, as many, if not all, FMS contain one-off elements. Teething problems are bound to occur and are likely to delay or even prevent full success of FMS implementations, irrespective of their causes.

Changes in market demand

Given the duration of the implementation process, changes in the marketplace are likely to occur and to have severe consequences for implementation success if they remain unobserved. As practical experience, illustrated

Variations in market demand between FMS specification and implementation may have significant effects. Company A encountered a 50% drop in market demand taking place during the implementation process and reacted by programming alternative products. However, the process planning department had not prepared itself properly for this contingency, by adopting process planning software (see Figure 5.11 for more details). Company B, whose FMS was laid out for the set-wise manufacture of all the major parts of a new family of pumps, achieved only 20% of anticipated output due to a much lower market demand than expected. However, the possibilities to use the FMS for alternative products were very limited as the system had been designed for a narrow envelope of components. Flexible manufacture of parts outside this envelope would require a virtually new FMS. Companies F and G encountered a higher demand than forecast, causing capacity constraint problems as the companies had aimed at a system utilization of about 70–80%. The same, though to a lesser degree, applied to company C which was looking for possibilities to manufacture high-volume parts conventionally, thus giving up the Just-in-Time delivery to assembly for these parts that the company had aimed at. The unexpectedly high market demand and the technical problems encountered forced company G to operate much more of its old production line than anticipated, even though this increased lead times and Work in Progress, due to problems of integration between existing facilities and the FMS. For much the same reason, company F also operated major parts of its old production lines, with similar effects as the ones encountered by company G. Moreover, annual volume of this company had reached a level at which, with hindsight, dedicated equipment had been a more cost-effective solution than flexible machinery. So, the company was considering the use of its FMS for manufacturing low-volume parts and start-up production of new variants only. Similar strategies will be applied to the FMS of cases C and D.

Figure 5.8 FMS in practice: the consequences of unobserved changes in market demand

in Figure 5.8, indicates, the volume flexibility of FMS is limited. Many companies still retain the conventional manufacturing wisdom that machine utilization should be as high as possible (Bolwijn and Kumpe, 1986). In itself this is not surprising, considering the high investment costs of FMS, which urge companies to make the best use of available capacity. However, this attitude incurs high risks if sales exceed available machining capacity. Increasing capacity by extending the system may be difficult and usually entails great expense. It is also difficult to integrate FMS technology and conventional equipment, without losing on flexibility and thus giving up at least part of the benefits pursued. There are not many possibilities to resolve this dilemma. Allowing for an investment in overcapacity could be one alternative. As another option, FMS could be used for the manufacture of low-volume parts. For example, FMS is a suitable alternative for start-up production during the introduction of new variants, and for products which are nearing the end of their life cycle (see Figure 5.9). During both stages, sales are less predictable and so are, in the early stages of growth, product modifications and (consequent) process modifications. Consequently, these stages require a flexible rather than a dedicated process. Dedicated facilities are much more suitable in the period of maturity, unless variants are many and volumes low. If such is the case, FMS may still be the most feasible alternative. However, the product and modification flexibility required for phasing in and phasing out parts depends not only on the technical features of FMS. Each change in the mix of parts produced by the system consumes a certain amount of process planning capacity needed for writing, testing and

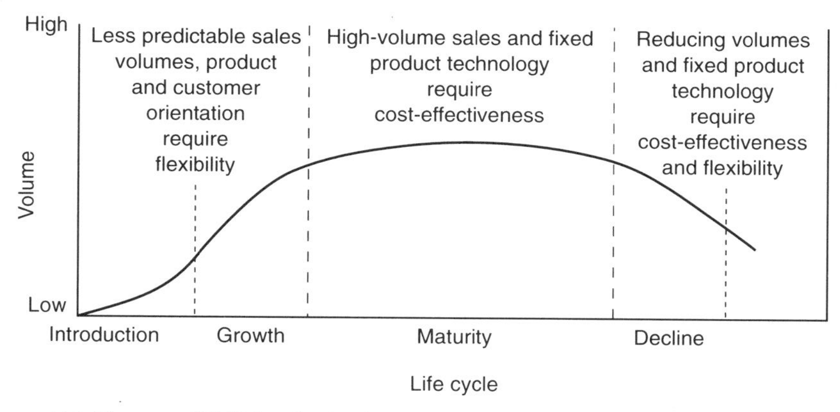

Figure 5.9 The use of FMS for phasing in and phasing out products

optimizing NC part programs and other software. Even more, if testing and optimizing need to be done on-line, that is at the system, machining time is consumed too, and production capacity lost. So, in order for this alternative to be feasible, additional measures need to be taken, such as the implementation of Computer-Aided Process Planning, to reduce process planning lead time and to enable off-line testing and optimization of software (see Chapter 7 in this volume). Problems similar to the ones described above may occur if the company attempts to react to slackening market demand by using the system to produce a wider range of parts. Then again, the company's product and modification flexibility play an important role.

The above strongly suggests that it is very important for FMS adopters to keep continuously informed of what is going on in the marketplace and to take appropriate action when needed. FMS implementation is a protracted process, lasting two to five years or even longer. During that period, many things may change, including mix and volume of demand, and the impact of that must be anticipated and taken into account when designing and implementing a flexible manufacturing organization using FMS.

ORGANIZATIONAL CONDITIONS

It is now generally recognized that many of the benefits of FMS arise from both the new type of organization which FMS requires and the technology itself. The benefits can be achieved only partially by 'just' implementing an FMS and using its characteristics. Certainly, the implementation of FMS does indeed contribute to the achievement of the benefits of FMS technology. Characteristics such as a certain level of integration and automation contribute to a reduction of direct labour. The flexibility incorporated in FMS leads to reduced set-up times which enables the adopter to produce smaller batches. Consequently, lead times and Work in Progress are reduced. These and other operational advantages may well improve the adopter's market responsiveness. Furthermore, the mere adoption of FMS

Processes
- undisturbed, smooth manufacture
- parts, tools, programs and other inputs available and loaded when needed
- preventive maintenance of machines, other system elements, software
- detailed process planning and NC part programming
- manufacturing based on realized assembly or customer orders, with the use of Just-in-Time principles
- optimal loading
- inspection of all inputs
- application of Total Quality Management concepts

People
- operators have integral knowledge of and skills in operating the FMS and solving any disturbances
- maintenance engineers have knowledge of FMS technology
- maintenance engineers are committed to and adequately skilled in tracing causes of failures
- people knowledgeable about and responsible for NC part program management
- people knowledgeable about and responsible for tool management
- process planners have NC part programming knowledge and skills
- production planners are able to develop optimal loading procedures
- people able to specify raw materials, programs and tools
- operators work according to operating procedures

Equipment
- diagnostic devices to trace causes of failures
- maintenance checklists
- stock of spare parts
- process planning software
- simulation software for developing optimal loading
- measuring tools for parts inspection, tool setting and assessing machine capabilities

Organizational arrangements
- multiskilled, semi-autonomous FMS team performing all routine operating, maintenance, production planning, program change and optimization, and quality control tasks
- rules and procedures covering these activities
- all-round supervisor acting as liaison with specialists
- (groups of) specialists in charge of non-routine maintenance, production planning, process planning and quality control tasks, respectively
- working groups of specialists in charge of solving problems requiring integrated solutions

Figure 5.10 Organizational conditions for the effective operation of FMS

will naturally increase the company's technological competence and improve the company's image as a technologically progressive firm.

However, the introduction of FMS causes a mismatch between technology and organization (Blumberg and Gerwin, 1984). Many structural, cultural and technological adaptations may be required to restore balance (Boer and Krabendam, 1991). The purpose of this section is to describe in more detail the flexible organization that is needed to be able to make full use of the flexibility of FMS (see Figure 5.10). A distinction is made between conditions related to the manufacturing, maintenance, process planning, production planning, and quality control processes; the people carrying out these processes; the equipment, other than the FMS itself, needed to perform these processes; and the organizational arrangements used to divide and co-ordinate the processes distinguished (Boer, 1991). The exact adaptations that must be made in order to achieve these conditions differ case by case, and depend on the FMS-related goals pursued by the company involved, the characteristics of the FMS purchased by that company, and the state of the company's manufacturing organization prior to the adoption decision. An account of the way the conditions were validated can be found in Krabbendam (1988) and Krabbendam and Boer (1989).

Processes

Reduced direct labour (especially if operations are unmanned) and increased utilization of capacity require an undisturbed input and manufacturing process, besides an integration and mechanization of operations, so that the transformation of raw material into finished parts can take place smoothly. The right amounts of raw material, pre-set tools and NC part programs need to be in place and loaded when operations are to start. These inputs, as well as the FMS transforming them, must meet the required quality standards.

Machine breakdowns need to be prevented, rather than remedied each time they occur. This calls for preventive rather than curative maintenance. Inspection lists specifying and scheduling regular inspections and preventive actions must be drawn up. If any breakdowns do occur, they need to be solved quickly. Furthermore, it should be recognized that cutting, fixturing, measuring and auxiliary tools, and other items of hardware, as well as NC part programs and other software, need maintenance as well.

Process planning needs to be done in much detail and with great accuracy, to ensure that NC part programs meet the required standards at once. Trial-and-error cycles of testing and adjusting NC part programs must be avoided as much as possible, to minimize loss of machining capacity and process planning lead time. The use of retrieval-type or generative process planning systems will be beneficial in achieving these prerequisites.

Lead time and Work in Progress are reduced through the integration of operations, but further reductions require shorter set-up times and smaller batch sizes. The former may be easily achieved using the technical flexibility of FMS, but the latter requires that the production planning process is aimed at scheduling the supply of raw material to manufacturing demands, and the manufacture of parts to be realized, rather than relying on forecast assembly orders. Centralized, computer-based production planning systems are not always suitable for scheduling small-batch manufacture on a day-to-day basis. The use of decentralized, for example *kanban*-driven, supply to and by the FMS is more promising. To achieve maximum output, techniques and procedures for optimal loading of the system must be developed.

Finally, quality control should be aimed at ensuring that the input to the system meets the required quality standards. Due to the level of automation of FMS, the quality of inputs, including raw material, tools and NC programs, must be predictable and constant. Just inspecting finished parts no longer suffices. The adopter should rather consider the application of (elements of) Total Quality Management (see Chapter 6).

People

In order to achieve short response times to operational problems, shopfloor workers need to be capable of handling a wide variety of problems that may disturb the manufacturing process. They need to be sufficiently trained and skilled to operate the different elements of the FMS and to resolve all kinds of problems that may occur in providing the system with raw material, software, and cutting, fixturing and measuring tools, and in the transformation process itself. Several procedures and rules must be designed and implemented, to standardize as much as possible the way the system is operated and controlled, and shopfloor workers must be trained to work according to these standards.

The complexity of FMS requires other and higher levels of knowledge than are usually available within companies, since operating and maintaining these advanced systems are more difficult compared with conventional machines, and require considerable competence in a wide range of mechanical, electronic and computer hardware and software technologies. Maintenance engineers need to be capable of operating special diagnostic devices required to identify sources of failures, wherever they occur. Furthermore, they need to be skilled in operating the equipment needed to measure machine geometry and other factors determining machining capabilities (see below). It is also important that maintenance people are trained in tracing causes of failures, rather than resolving symptoms. As cutting tools, fixtures, part programs and other software are critical components, people should be appointed and made responsible for the management and maintenance of tools and software.

Process planners need sufficient knowledge about FMS technology and must be skilled in planning and programming all the wide variety of operations to be performed by the system. Furthermore, they must be trained in using Computer-Aided Process Planning systems.

Production planners must be sufficiently skilled to be able to design and implement optimal loading procedures, possibly with the assistance of process engineers.

Finally, quality people must be trained to design specifications for raw materials and for the NC part programs and tools, which are needed to achieve the required constant and predictable input to and output from the system.

Equipment

In order to keep the FMS functioning effectively, additional pieces of equipment may need to be purchased. To start with, the complexity of FMS is so great that special diagnostic devices are required to detect causes of software and control problems. Furthermore, critical spare parts must be held on stock. Computer-aided Process Planning tools are needed, which require special software and, possibly, hardware. The production planners involved in designing optimal loading procedures, must have the required software and data-processing capacities at their disposal. Quality controllers may need advanced and quick measuring equipment, to keep up with the system's output rate. Furthermore, equipment is required to ensure the system's capabilities, including the machining centres, tools and fixtures.

Organizational arrangements

The implementation of all organizational conditions needed to obtain the benefits of FMS requires the adopter to reconsider its working organization. Basically there are three alternatives. In the first alternative, each of the functional departments involved controls only those aspects for which it is responsible. However, this purely functional form of organization is too expensive and time consuming, considering the promises, and in particular the flexibility, of FMS. If any problems occur, too much time will be lost in activating maintenance, process planning, or other specialist departments. The other extreme is the self-contained group. The drawback of this is that too many highly skilled people are spending too much of their time performing relatively routine activities which are regulated by the system. In the long run, this situation will become unattractive to them.

The most feasible alternative is a flexible organization based on a semi-autonomous FMS group of well-trained, multiskilled operators who are, and feel responsible for, carrying out all of the more or less routine operating, maintenance, production planning, and quality control tasks, as well as program changes and optimizations. Rules and procedures regulating these activities are designed and implemented by specialists in these fields. The group is managed by a supervisor whose main role is to act as liaison between the FMS group and the specialists. In addition, a multifunctional working group, staffed by different specialists, supports the FMS group by providing solutions to problems requiring an integrated approach, such as standardizing tools, specifying raw material and finished parts, selecting suppliers, and phasing in new parts. Furthermore, this working group may assist the operators in looking for possibilities for the continuous improvement of the system's performance.

The implementation of an FMS and all the organizational conditions outlined before is a complex process. In the next section some of the key organizational issues affecting the success of that manufacturing innovation process are highlighted.

IMPLEMENTATION MANAGEMENT

The adoption and implementation of FMS is a protracted process. Several years elapse between the first idea to replace present equipment and the installation of the new system. During the initial stages of the innovation

In the majority of the cases, the FMS was not used to resolve a strategic bottleneck, except for company A whose major objective was to reduce delivery time of products that were made to order. Unfortunately, however, that company achieved only 50% of anticipated output as its market collapsed during the implementation process. In order to increase the system's utilization the company decided to try and find alternative products by subcontracting overcapacity. As a result, however, the mix of parts produced by the system changed considerably. Writing new NC part programs and proving them had to be done at the system which not only consumed considerable machining capacity but also virtually all of the reduction of manufacturing lead time achieved. So, the bottleneck moved, from manufacturing lead time to process planning lead time.

The company did not take much time in deciding to adopt process planning software needed not only to reduce process planning lead time but also to enable the off-line writing, testing and optimizing of NC part programs needed to increase the system's availability.

Figure 5.11 FMS in practice: moving bottlenecks

process, a wide variety of interdependent decisions concerning organizational and technical aspects must be made, which have considerable influence on the eventual performance of the system. However, the trial point of these decisions lies at least one or two years ahead. Furthermore, during the process several market-related factors may change, and this may have a devastating impact on the success of the innovation. In effect, a lot of uncertainty is involved in the decision to purchase FMS. In order to increase their ability to cope with this factor and to increase the likelihood of the FMS contributing to the goals pursued, FMS adopters should take the following into consideration.

Firstly, as indicated before, they should realize that implementing an FMS and using its technical flexibility and integration, mechanization and reprogrammable automation of operations, does not automatically lead to *operational advantages*. In order to fully achieve these benefits, many organizational adaptations in production, process planning, production planning, quality control and maintenance are required, which depend on the goals set and the initial state of the company's manufacturing organization. Also, FMS adopters will not automatically achieve *market advantages* by installing an FMS to replace less flexible equipment. The achievement of these advantages does depend on the organizational conditions indicated previously but, in addition, the system must be installed in a 'strategic' bottleneck, that is a part of the organization reducing the company's success in the marketplace (see Figure 5.11). In other words, if other organizational factors, whether they concern processes, people, equipment or organizational arrangements, are reducing product quality or flexibility, delivery time, delivery reliability or cost-effectiveness, the FMS will not contribute to the company's business success, or only partially. Then the use of other manufacturing technologies, operations management systems, forms of organization, or other mechanisms aimed at increasing market responsiveness may be much more promising.

Secondly, future FMS adopters should be aware of the inevitability of changes taking place during and after the innovation process. This means

None of the companies perceived the implementation of FMS as a manufacturing innovation problem. This perception led the companies to adopt a one-sided, technical approach to resolving the implementation problem. FMS technology was considered the panacea for all companies' problems. However, the majority of the companies underestimated the diversity of the activities to be performed in order to get the system operating effectively, and decided to have their process planners handle the innovation. The other core functions were either not involved or to a very limited extent. In all the companies the manufacturing department had a say as far as the selection of and the training programme given to the future FMS operators were concerned. The production planning department played no significant role in half of the cases. Usually, the maintenance function was involved to the extent that they were requested to assist with the preparation of the FMS site and the installation of the system. In cases B and C only, the quality control department was actively involved prior to installation. In cases C and D the production director was frequently involved in the project. In case B the project was led by the general manager. In the other companies, the role of general management was largely restricted to approving the investment proposal.

The companies paid insufficient attention to the exchange of information between functions. Most information was exchanged using established, formal channels of communication. All companies established a project team for the duration of the process. However, dominated by the process planning function these teams focused on technology, rather than the design and implementation of the techno-organizational solution needed to resolve the innovation problem. In doing so, the process planners actually performed an excellent job. Not only did they specify the FMS, they also designed many of the required organizational adaptations. However, issues falling outside their scope could not be recognized or handled adequately prior to the installation of the system simply because they were insufficiently skilled to appreciate all of the reorganization required. Failing to understand that the technical characteristics of FMS would contribute only partially to the benefits pursued, they put insufficient effort into translating these benefits into an adequate manufacturing organization. In none of the cases other than the most obvious, organizational specifications were added to the technical ones, and this happened only after the adoption decision had been made.

Only after installation of the system, with technical problems abundant and operational experience increasing, did the companies start the implementation of additional conditions. Some companies decided to establish temporal linkages among the departments affected by the FMS. The other companies kept relying on their existing structure which, however, was not very suitable for handling the wide diversity of problems encountered. Consequently, the additional adaptations were the result of trial-and-error learning by separate functions, rather than a concerted effort. As a result, all companies but company C had to spend more time and effort than strictly needed to achieve the innovation pursued. However, none of the companies had built much time or financial slack into their project. On the contrary, all the companies had attached short-term financial criteria to the investment and put the project under time pressure.

As a result, none of the companies achieved all of the FMS benefits pursued at the required time. At the end of the research, companies B and C were the most successful firms in the sample, though neither of them had achieved all the benefits pursued. Companies A, D and F were about half-way to goal achievement, whereas companies E and G still had a long way to go. Consequently, financial goals were automatically reduced.

Figure 5.12 FMS in practice: implementation management

that they should put a great deal of effort and time into formulating realistic goals and designing the organizational conditions and technical specifications contributing to the achievement of the goals set. Furthermore, they should be prepared to change these factors if circumstances give rise to such a decision. A very effective, if not the best, way of coping with changing market demands, is to allow for slack flexibility in the form of, for example, overcapacity, the use of multiskilled operators and the establishment of multifunctional teams.

A wide range of functions need to be involved, organically co-operating to prepare the investment decision and determine the organizational adaptations required for the effective operation of FMS. Among the core functions that need to be involved in deciding whether or not to adopt FMS, and in setting realistic and achievable goals, are production, process planning, maintenance, quality control, and production planning and scheduling (see Figure 5.12). Furthermore, 'boundary spanners' such as product design, purchasing, marketing and sales may need to be involved as their knowledge of the company's product and market is needed to assess what manufacturing system is most appropriate for the company to improve its market performance. Conversely, the new system may create opportunities not previously possible. Then the company's market strategy may need adjustment to its new manufacturing potential, a decision that requires the involvement of the 'boundary spanners'.

Anyway, throughout the design, production, installation and operation of the new technology and its organizational environment, these functions must interact to define and solve the real innovation problem. And this problem is much farther reaching than the installation of just another piece of equipment. Essentially it requires the definition of the technical and organizational conditions described before, implementing these conditions, and maintaining adequate standards after installation.

All this means that the successful implementation and operation of FMS requires an integrated approach to the management of innovation. Clearly, such an approach requires a proactive organization in which capital investments are based on a comprehensive Manufacturing Strategy rather than just technical or operational needs.

In order for the Flexible Manufacturing System and organization to be implemented successfully, higher management is to play an active role, championing, sponsoring and, if necessary, even organizing the manufacturing innovation process. They cannot limit themselves to making go/no-go decisions, based simply on quantifiable operational advantages assessed by means of questionable justification methods. Instead, it is their responsibility to decide on the role that combinations of FMS, or other types of Advanced Manufacturing Technology, novel management techniques and modern forms of organization are to play in the company's Manufacturing Strategy. Furthermore, manufacturing innovation requires a flexible organization capable not only of generating and processing a wide variety of information, ideas and potential solutions, but also of learning from the implementation process and the operation of new manufacturing systems. So, a crucial factor determining success or failure is the creation of sufficient lateral communication between the functions indicated before. As many FMS adopters are typical examples of machine bureaucracies, general management are often the only ones able to resource the innovation process appropriately and to break down the boundaries between functional departments. Only then will organizations experience fewer of the problems frequently encountered during and after the installation of FMS.

A major spin-off of such an approach is increased knowledge of how to tackle strategy-driven manufacturing innovation. And with competition and market pressures increasing, on the one hand, and an ever-growing number of new technologies and novel management techniques becoming available, on the other, manufacturing innovations are going to abound in many industries.

CONCLUSION

FMS technology promises much in terms of improved manufacturing and consequent market success. This chapter, however, has shown that the extent and speed of goal achievement depend a great deal upon company-wide involvement in the design and management of organizational innovation.

REFERENCES

Bessant, J. R. (1989) Flexible manufacturing; yesterday, today, tomorrow, in H. Bolk, H.-U. Förster and W. Haywood (eds.) *Proceedings of the International Conference on Implementing Flexible Manufacturing*, Amsterdam, 25–27 January.

Bessant, J. and Haywood, B. (1986) Flexibility in manufacturing systems, *Omega*, Vol. 14, no. 6, pp. 465–73.

Bessant, J. and Haywood, B. (1988) Flexible manufacturing in Europe, *European Management Journal*, Vol. 6, no. 2, pp. 139–42.

Blumberg, M. and Gerwin, D. (1984) Coping with advanced manufacturing technology, *Journal of Occupational Behaviour*, Vol. 5, pp. 113–30.

Boer, H. (1991) *Organising Innovative Manufacturing Systems*, Gower, Aldershot.

Boer, H., Hill, M. R. and Krabbendam, J. J. (1990) FMS implementation management: promise and performance, *International Journal of Operations and Production Management*, Vol. 10, no. 1, pp. 5–20.

Boer, H. and Krabbendam, J. J. (1991) Organizing for market-oriented manufacture. Paper presented at the POMS Conference, New York, 11–13 November.

Bolwijn, P. T. and Kumpe, T. (1986) Toward the factory of the future, *The McKinsey Quarterly*, Spring, pp. 40–9.

Braun, E. (1981) Constellations for manufacturing innovation, *Omega*, Vol. 9, no. 2, pp. 247–53.

Charles Stark Draper Laboratory (1984) *Flexible Manufacturing Systems Handbook*, Noyes Publications, Park Ridge, NJ.

Finnie, J. (1988) The role of financial appraisal in decisions to acquire advanced manufacturing technology, *Accounting and Business Research*, Vol. 18, no. 70, pp. 133–9.

Gerwin, D. (1988) A theory of innovation processes for computer-aided manufacturing technology, *IEEE Transactions on Engineering Management*, Vol. 35, no. 2, pp. 90–100.

Groover, M. P. and Zimmers, Jr, E. W. (1984) *CAD/CAM: Computer-Aided Design and Manufacturing*, Prentice-Hall, Englewood Cliffs, NJ.
Hartley, J. (1984) *FMS at Work*, IFS (Publications) Ltd, Bedford.
Haywood, B. and Bessant, J. (1987) Flexible Manufacturing Systems and the small to medium-sized firm. Occasional Paper No. 2, Innovation Research Group, Brighton Polytechnic.
Hill, M. R. (1985) FMS management – the scope for further research, *International Journal of Operations and Production Management*, Vol. 5, no. 3, pp. 5–20.
Hill, T. J. (1989) *Manufacturing Strategy: Text and Cases*, Irwin, Homewood, Ill.
Jaikumar, R. (1986) Post-industrial manufacturing, *Harvard Business Review*, November–December, pp. 69–76.
Kaplan, R. S. (1986) Must CIM be justified by faith alone? *Harvard Business Review*, March–April, pp. 87–93.
Krabbendam, J. J. (1988) *Nieuwe Technologieën en Organisatorische Maatregelen*, University of Twente, School of Management Studies, Enschede.
Krabbendam, J. J. and Boer, H. (1989) Anticipating and managing organisational measures for the implementation of new technologies: the case of FMS. Proceedings of the 2nd International Production Management Conference on Management and New Production Systems, Fontainebleau, France, 13–14 March.
Lunt, P. J. and Barclay, I. (1987) The successful management of the introduction and operation of advanced manufacturing technology. Proceedings of the Annual Meeting of the American Society for Engineering Management.
Macbeth, D. K. (1985) Flexible manufacturing – the hope for European industry, *European Management Journal*, Vol. 3, no. 1, pp. 27–32.
Primrose, P. L. and Leonard, R. (1985) Evaluating the 'intangible' benefits of flexible manufacturing systems by use of discounted cash flow algorithms within a comprehensive computer program, *Proceedings of the Institute of Mechanical Engineers*, Vol. 19, no. B1, pp. 23–8.
Tombak, M. and De Meyer, A. (1988) Flexibility and FMS: an empirical analysis, *IEEE Transactions on Engineering Management*, Vol. 35, no. 2, pp. 101–7.
Voss, C. A. (1988) Implementation: a key issue in manufacturing technology. The need for a field of study, *Research Policy*, Vol. 17, pp. 55–63.

6

TOTAL QUALITY MANAGEMENT

Patrick Dawson

Total Quality Management (TQM) is a widely used strategy for increasing organizational flexibility and employee commitment to change. It originated in America, was developed in Japan and is now forming a central part of strategic decision-making in companies in North America, Asia, Europe and Australia and New Zealand. Unlike quality circles, TQM is not viewed as a purely operational technique for improving shopfloor participation but, rather, is seen as a method for integrating programmes of operational control with the strategic management of a company. In this way, TQM has developed from a series of specific techniques, such as the use of Statistical Process Control, to the complete management of intra-company relations and the creation of policies and practices aimed at securing employee involvement (Dawson, 1994). The 1990s have seen a movement away from Total Quality Control (TQC) which, following Feigenbaum (1961), had a strong association with production, towards the management of organizational culture and the enlisting of all parts of an organization in the systematic effort for quality (Hames, 1991). The emphasis is now on the involvement of all levels of employees, in all different types of organization, for the purpose of continuously refining the process of service and product delivery with the aim of improving operational efficiencies and increasing employee commitment at work (Macdonald and Piggott, 1990). Quality issues are no longer restricted to production departments or manufacturing companies, but are the concern of the whole gamut of public and private enterprises who wish to enhance their system of employee relations, reduce operational inefficiencies and increase their competitive position (Ballantyne, 1992; Dawson and Patrickson, 1991).

In a recent article, Palmer and Saunders have discussed the significance of quality management programmes for Human Resource Management (HRM) and claim that there is a need for more systematic analysis and investigation to separate 'real outcomes from the extravagant claims that go with each attempt by management consultants to promote new forms of managerial prescription' (Palmer and Saunders, 1992: 77). At present, there is a general absence of detailed empirical data on the implications of TQM for employment relations at work. This is perhaps surprising given the wide organizational uptake of TQM and the supporting plethora of prescriptive literature which advocates the replacement of individual-based work practices with systems which encourage teamwork and employee participation on the shopfloor (see for example, Berry, 1991). Furthermore, in the search for quality, corporations are not focusing solely on intra-company relations, but also on the development of collaborative inter-organizational relations with their major customers and suppliers (Oakland, 1989: 4). In short, TQM is increasingly being used as a cultural control strategy which goes beyond the shopfloor to incorporate strategic Human Resource Management and inter-company relationships. It is a new and emerging manufacturing and service strategy which remains clouded by the hype and gloss of neat prescriptive *post hoc* rationalizations and market-driven consultant packages and, as such, is an area in need of further detailed investigation and longitudinal critical appraisal.

This chapter sets out to clarify some of the common confusions surrounding the debate on quality management through identifying some of the main characteristics of TQM and then describing their implementation within a manufacturing context. A combination of technical and social factors is shown to colour the nature of the TQM experience and influence the pace, pattern and consequence of change for company employees. The process and outcomes of this New Wave Manufacturing strategy on employment relations and work arrangements are examined in three main sections. In section one, a brief historical overview of the emergence of TQM is provided in order to identify the nature of the conceptual confusion which surrounds TQM. This is followed by a description of the main characteristics of TQM programmes. Section two then analyses the process of managing change and the practice of establishing TQM within a manufacturing environment. The case of Pirelli is used to illustrate the reality of TQM and calls into question the total nature of TQM participation. In section three, four general lessons on the management of TQM are outlined and the extent to which contextual factors shape the shopfloor characteristics of TQM is discussed. The section concludes with an appraisal on the implications of TQM for HRM, in which it is suggested that many of the current principles of TQM contrast and conflict with existing HRM practices.

THE CONCEPT OF TOTAL QUALITY MANAGEMENT

The historical emergence of Total Quality Management

There remains considerable confusion over what constitutes TQM. In part, this can be explained by the history of quality management and the development of quality programmes which have broadly moved from the establishment of principles of quality assurance, to issues of quality control, and finally, to incorporate the more behavioural and attitudinal aspects of quality which are intended to be encompassed in the term 'Total Quality Management'. Although some people continue to use the terms Total Quality Control (TQC) and Total Quality Management (TQM) interchangeably, there is an important distinction between them.

Total Quality Control (TQC) originated from developments in quality control (QC) which were based on the development of statistical techniques for measuring actual quality against established quality standards. In 1924, Shewhart (who worked in the quality assurance (QA) department of Bell Laboratories in the USA) developed a series of statistical control charts which were able to detect changes in the variability of a production process before defects were produced, rather than separating defects from acceptable components after production (Allan, 1991: 24). These early findings were further developed by Deming, Juran and Feigenbaum, but were not widely taken up by industry until the Second World war. In the post-war period, the use of these quality control methods was limited with American industry placing an emphasis on quantity rather than quality (Walton, 1986: 8–9) and British management being reluctant to adopt and adapt these techniques.

Armand Feigenbaum continued to develop the principle of QC and, in an article in the *Harvard Business Review* in 1956, used the term 'Total Quality Control' to convey the view that quality is the responsibility of all groups within an organization (Campell and Davies, 1992: 4). He claimed that everybody should be involved in the process of satisfying customer requirements, rather than quality being the preserve of a small group of specialists in QA departments (Feigenbaum, 1961). In this sense, the development of TQC has historical connections both with QA and the more recent developments in TQM. The main difference with QA is that TQC is about total employee commitment in the systematic effort to achieve quality.

In the case of TQM, there are a number of similarities with TQC and many of the principles developed under TQC are readily transportable to companies operating TQM programmes. The main difference is one of industry emphasis, that is, TQC programmes have tended to focus on the use of statistical techniques in manufacturing industries with little attention being given to service-based companies. In contrast, TQM is seen to have a broader application in being based on a general philosophy for involving employees in the pursuit of quality objectives. It also differs in so far as TQM programmes tend to place greater significance on techniques for achieving an increase in employee commitment and the development of high-trust relationships. The non-tangible and cultural elements become the keystone to strategic change, with the operational statistical techniques and group problem-solving forums being the method for gaining employee involvement (Dawson and Palmer, 1994). In contrast to both quality circles and quality control, the focus with TQM has shifted from *operational techniques* for improving quality or employee involvement, towards the *strategic management* of attitudes and behaviours and the development of collaborative intra-company and inter-organizational relations. In other words, cultural change strategies are combined with operational techniques for enhancing customer–supplier involvement and commitment.

Common characteristics of Total Quality Management programmes

There are numerous definitions of TQM and considerable disagreement on what constitutes a TQM programme. For some writers, TQC is synonymous with TQM (Blakemore, 1989); for others TQM has been redefined as Total Quality Service (TQS) to account for the special characteristics of service organizations (Albrecht, 1992; Berry and Parasuraman, 1991; Zeithaml, Parasuraman and Berry, 1990), and for others TQM is seen to be composed of a distinct set of principles (see, for example, Fox, 1991; Tenner and DeToro, 1992). What is common among these apparently different approaches is their attempt to delineate a unique methodology for the successful implementation of TQM in manufacturing and/or service organizations. The tendency to differentiate TQM programmes is not surprising when one considers that the published material on TQM has generally been written by consultants with a vested interest in promoting their own distinct ideas (see also Dobyns and Crawford-Mason, 1991: 20–1).

The problem of defining a total approach to quality also stems from the development of TQM as a more general philosophy of change which, whilst having its roots in many of the principles of Total Quality Control, now encompasses a broader and more strategically oriented change management model aimed at revising existing organizational attitudes and beliefs systems. This cultural dimension of TQM is often heralded as the 'unquantifiable heart' of modern quality programmes and yet has proven very difficult to delineate and define. The vested interests of competing consultant groups, the incorporation of service industries into quality programmes, the long-term nature of TQM change strategies, and the growth in the application of quality management to include cultural and attitudinal change – as well as the more conventional application of Statistical Process Control techniques to shopfloor operations – all render it difficult to construct a single definition for TQM. Nevertheless, there are a number of common characteristics which tend to be present in discussions and debates on TQM and these will serve as a useful starting point for explaining the core principles of TQM programmes (see also, Dawson and Palmer, 1993: 116–23).

The first key feature of TQM is that it is a *total management approach* to quality improvement which involves *every employee* of the company, as well as internal and external operating practices and customer–supplier relations. This holistic approach to quality management differentiates it from historically earlier attempts at quality control (through quality assurance departments) and employee involvement programmes (through the use of quality circles on the shopfloor). It is a total management philosophy based on the view that change is a necessary and natural requirement of organizations wishing to keep pace with dynamic external business market environments and continually improve existing operating systems. The total approach requires the backing and commitment of senior executive management (who are expected to endorse strongly the programme and align themselves with TQM objectives), other managerial groups, supervisors and shopfloor and branch-level personnel. In short, it is a philosophy of change based on senior executive commitment to total employee involvement in the group problem-solving of processes issues.

The second main characteristic of TQM is its emphasis on *continuous improvement* or incrementalism. Unlike many operational change programmes which aim at restructuring an organization and then stabilizing operating systems through the use of clearly prescribed sets of procedures, TQM programmes are based on institutionalizing a system of continuous process improvement. The principle of ongoing change in process operations is established as normal operating practice. Minor developments and changes in process operations are no longer viewed as a threat to established working relationships but, rather, become an employee expectation and form part of a new routinized way of working based on incremental change and process improvement. In Japan the term *kaizen* has been used to refer to the system which supports gradual, unending improvement for achieving ever-higher standards in meeting changing customer and market requirements. In the words of Masaaki Imai,

> The essence of *kaizen* is simple and straightforward: *kaizen* means improvement. Moreover, *kaizen* means ongoing improvement involving everyone, including both managers and workers. The *kaizen* philosophy assumes that our way of life – be it our working life, our social life, or our home life – deserves to be constantly improved.
>
> (Imai, 1986: 3)

Kaizen (and the principle of incremental innovation in process operations) has been the basis from which many Japanese systems have been developed. It is an essential characteristic of TQM programmes which seek to make top management and all other employees '*kaizen* conscious' in being committed to continuous improvement in operating practices.

This leads to the third element of TQM programmes, namely, the *application of appropriate quality control techniques*. Although some commentators have argued that 'sustained quality improvement without the use of quality measurement techniques is impossible' (Fox, 1991: 7), with TQM the emphasis is not on systematic measuring techniques *per se*, but on the use of statistical methods to support the objective of continuous process improvement. Traditionally, quality control has been the concern of quality inspectors who monitor and evaluate finished products against a set of requirement specifications. Those products which meet the specifications are sent on to customers and those which fail are either scrapped or reworked. In practice, there may also be a degree of adjustment and revision to product quality standards in response to such factors as: market demand, type of customer and perceptions of customer dissatisfaction. Under these conventional systems problems of poor quality are addressed with the end product or service rather than at the stage in which the fault in producing the good or service occurred. Under a TQM system, the aim is to identify and eliminate process problems at their source. This is achieved in groups (see below) through the use of various statistical tools, which are used to provide a simple way for monitoring and evaluating process operations. Some of the more common statistical techniques include: control charts, statistical measures and sampling, design of experiments, evolutionary operation, and Pareto analysis (see Brocka and Brocka, 1992: 270–304). For example, a shopfloor operator may be trained in how to take a sample of items from a production line and then to measure and record these items on a control chart. The charts would normally have two lines which set the upper and lower control limits, indicating whether the process is in control (that is, within these parameters) or whether corrective action is required.

The fourth and related characteristic of TQM is the use of *group problem-solving techniques* to process operations. There are a number of different techniques which are used to facilitate team problem-solving activities (see, for example, McConnell, 1988: 3–134). For example, brainstorming is a common tool used during the initial stage of identifying a process which requires attention (Blakemore, 1989: 73). The aim is to maximize the number of suggestions through encouraging the participation of all members and ensuring that no evaluations are made during the brainstorming session. Following this, the team may identify a few important items and eventually a decision will be made about which process should be the focus of the group's attention. Flow charts may then be used to increase the groups understanding of the process in question. They are mainly used to help break down problems into smaller parts and simplify complex structures and inter-relationships. The next step involves measuring the process through collecting various data which can be represented on simple run charts and histograms. The group may then decide to run a further brainstorming session to identify the range of problems generated by the data (Dawson and Palmer, 1993: 120–21).

A fifth major component of TQM programmes is the *focus on 'internal' and 'external' customer–supplier relations*. The internal customer is the next person or group in the process of manufacturing goods or providing services. Apart from those involved in external customer–supplier relations, everyone should have an internal customer and under TQM should aim to meet the requirements and expectations of that customer. In order to meet customer needs it is first necessary to clarify what those needs are and this requires open and regular communication between the various customer–supplier groups. In stressing the importance of these internal relationships, TQM highlights the importance of teamwork and communication, and illustrates how external customers' requirements cannot be achieved if each output passed between employees within the company is deficient (Tenner and DeToro, 1992: 52). The concept that quality is defined by the customer is central to TQM and whilst it is expected that internal suppliers focus on meeting the needs of their immediate customers, an overview must also be maintained of the expectations and requirements of the external customer. In some cases customer requirements may be vague and in others they may be unrealistic; in either event TQM provides the vehicle for developing stronger links with customers and clarifying requirements. The concept of internal as well as external customers therefore encourages not only greater inter- and intra-organizational communication, but also wider customer–supplier participation in decision-making about process improvements to meet product and service requirements.

The sixth element associated with the principles of TQM centres on the *building of high-trust relationships and the development of non-adversarial systems of industrial relations*. Through the development of stronger internal relationships and the implementation of policies which seek to devolve greater decision-making to operative groups, TQM programmes aim to create and sustain high-trust relationships and employee co-operation on the shopfloor. The communication process is the central means to building trust between senior management, middle management and other employees within the organization. The objective is to bring about a shift in attitudes and remodel the traditional adversarial system of industrial relations through a more open and participative management approach which actively seeks and places a premium on the knowledge and experience of all employees. To put it simply, change is initiated through TQM teams and supported by management, rather than being imposed by management on the workforce.

Finally, it is important to note that organizations operating under a management philosophy of TQM would normally have introduced new organizational arrangements to accommodate such a change. The organizational structure for TQM is likely to vary between organizations, although it would generally mirror rather than supplant existing reporting and command structures. Typically, there would be a steering committee formed by the executive of the company who would set policies and quality programme objectives, recommend and approve TQM quality groups, monitor and evaluate progress, and ensure that the TQM programme fitted with the overall strategic direction of the company. Below the steering committee there may be a number of quality groups located in different plants or assigned to different functions, such as manufacturing, research and development, finance and administration, and sales and marketing (see also, Fox, 1991: 89). At the branch or shopfloor level there would be a number of quality teams all tackling particular problems, such as machine downtime in manufacturing. In addition to this basic structure, there would normally be a TQM co-ordinator with responsibility for directing and evaluating the implementation of quality programmes and recommending new TQM initiatives. The TQM co-ordinator would also provide support to the various quality groups and assist in the training, development and education of employees.

These major characteristics of TQM programmes highlight the importance of employee involvement and teamwork to the success of this new management philosophy which places a premium on participation and continuous process improvement. The process of managing the introduction of these principles into a manufacturing organization and the implications of these changes for employment relations are now discussed in a case study examination of Pirelli Cables.

THE PRACTICE OF TOTAL QUALITY MANAGEMENT

A case study of Pirelli Cables

In 1992 Pirelli Cables Australia Limited (PCAL) consisted of three manufacturing sites and eleven sales branches (including one in New Zealand). The Minto site is the newest plant; it is located in an outer southern suburb of Sydney and specializes in power and building wire cables. The oldest plant is situated in a northern Sydney suburb, where production operations centre on the manufacture of telephone cables. Two plants are located in South Australia at the Adelaide manufacturing site; namely a cable manufacturing plant and a cable processing plant. The former comprises comparatively complex equipment for the manufacture of single- and multi-core flexible cables. The extrusion equipment is old and therefore requires routine maintenance. Of the 47 operators 46 are male and the equipment is operated on a three-shift basis. In contrast, the cable processing plant is predominantly female (90 of the 106 employees) and is based largely around labour-intensive repetitive manual processing operations. This plant processes cable manufactured by the cable plant in order to service customer requirements (for example, in the type of plug and length of cable). However, Pirelli Cables has little worldwide expertise in the area of processing (cable manufacturing predominates) and, consequently, the cable processing plant is largely isolated from the other Australian manufacturing operations.

Establishing TQM: a management approach for continuous improvement

The starting point for the introduction of TQM at the Adelaide manufacturing site began in 1989. There was a long period of discussion with the overall site manager (who was part of the senior executive group) and the other executive members in New South Wales. Within PCAL as a whole, the TQM project commenced on 6 March 1989 and the plan was to implement TQM throughout their Australian manufacturing operations in three stages. The Minto factory was classified as stage one, followed by the Northern Sydney plant and, finally, Camden Park in Adelaide. A steering committee was formed from PCAL senior executives, who were trained by Blakemore Consulting in the principles and philosophy of TQM. The managing director acted as the chairman of the steering committee, and other key personnel were trained as facilitators in preparation for project team formation. At the Minto site in New South Wales, four project teams were formed initially, with each team member being trained by the consultant group and supplied with training manuals. Another five teams were added during 1989 and by the end of 1990 nine teams were actively participating in the Minto TQM programme. The teams covered a number of areas including: labour turnover, customer delays, customer complaints, accounts lead time, cable defects and scrap reduction.

Stage two of the TQM project commenced on 14 November 1989 at the Dee Why site in northern Sydney. Initially, four project teams were formed involving 38 employees. Stage three was expected to commence shortly afterwards, but due to the airline strike there was a considerable delay, as the site manager commented:

> It just so happened that it lined up exactly with the airline strike and so we didn't get going until about April or May the following year [1990]. Then we had two days of training with selected groups of 20 on each day. Basically we went through all the concepts of TQM and the statistics and so forth. Each group had one day, so there were 20 people one day and 20 the next. So we covered 40 to 50 odd people. These were people that we thought would be put into teams immediately or some time in the future.
>
> (senior executive interviews, 1991)

Stage three of the TQM project commenced in Adelaide on 14 May 1990 and three project teams were formed dealing with maintenance downtime, machine improvement, and scrap reduction. By the end of 1990, approximately 150 Pirelli employees had been involved in some form of TQM training during the initial three stages of TQM implementation.

In the first 1991 steering committee meeting held at the Minto headquarters on 20 February (in which the author was an observer), it was decided that TQM team membership would be restricted to a maximum of eight personnel, as the 1990 teams of ten had proven 'too unwieldy'. It was also recommended that employees should only be allowed to attend two one-hour meetings per week (previously some employees had become members of four teams and were spending four hours per week in TQM meetings). The reasons for this restriction were twofold; first, to discourage TQM 'freeloaders' who appeared active on paper but were not making significant contributions either to the teams or production; second, to ensure that there were always enough personnel on the shopfloor to minimize disruptions to production caused by problems of work allocation associated with a prolonged reduction of personnel.

In discussing the future strategic direction of TQM, it was noted that it was very difficult to allocate costs and cost savings associated with the TQM project. Although it was estimated that cost savings of around $1,107,000 per annum would be achieved as a result of improvements and action plans of TQM team participation, no estimate of the total cost of introducing TQM was available. For many of the executives, the emphasis on dollar returns missed a key ingredient of TQM; namely, that it was about gaining employee commitment to the continual search for quality improvements. As the manager of the Adelaide operations commented,

> I have a lot of trouble quantifying what the savings are off the top. They are all intangibles, and I believe intangibles far outweigh what you are paying, but I would have a lot of trouble putting dollars down. I could probably do it for two of the teams, but then it might have happened anyway. You can't say that it is totally because of TQM, it might have been. I would have no trouble arguing the case, whether I could actually put dollars to it, I don't know.
>
> (senior executive interviews, 1991)

One of the major strategic objectives of TQM was to bring about a 'culture of change' and initiate a shift in employee attitudes through the development of high-trust relationships and a commitment to continuous process improvement. TQM was seen to provide a total management approach to securing employee involvement and creating a more collaborative system of industrial relations.

Total employee involvement and the system of industrial relations

Employees who had been involved in TQM activities claimed that TQM was of benefit both to shopfloor personnel and to the company as a whole. The benefits to employees were the possibility of improving working conditions and reducing some of the stress and frustration associated with rework caused by minor problems; whereas, the potential advantage for the company was increased efficiency rates and reduced scrap. For example, one of the biggest changes to shopfloor operations has been the increase in daily interaction between all employees which has been supported by TQM activities. TQM is seen to have facilitated greater communication between shopfloor personnel and improved employee understanding of the processes involved in manufacturing. There is now a greater willingness among operators to help each other out if there are problems in particular areas. Furthermore, union representatives supported these changes and did not view TQM as a threat to either job security or shopfloor employment conditions. The National Union of Workers (NUW) had been informed about the decision to introduce TQM and had lodged no concerns with management. For example, on being asked to describe the union view on TQM a NUW shop steward indicated that they did not view TQM as a priority concern and that there had been no involvement by the various shop stewards other than as being members on some of the TQM teams. Finally, when asked about the union's views on management, the shop steward replied, 'Well management, from a union point of view, I have no problems with them. There is quite a good relationship between the union and management here' (shop steward, 1991).

The evaluation of TQM by union representatives was favourable and supportive of the more general move towards greater teamwork and collaboration on the shopfloor. In practice, however, the context within which change was being introduced acted as a major constraint on the development of a more harmonious system of customer–supplier relations at plant level.

The context of change and the issue of customer–supplier relations

In examining the management of TQM and observing operations at Pirelli Adelaide, the author recorded the very different operating environments in the cable manufacturing and processing plants. In the cable manufacturing plant there was a type of 'machine-dominated' working environment, in the sense that comparatively large and heavy machinery was being used to draw down copper in the manufacture of cables. The plant was fairly dirty, was noisier than the processing plant, and had that 'engineering smell' associated with oil and grease. Although the plant is not staffed solely by male employees, on first impressions the large number of male operators, combined with the sound and aroma of 'machines-in-motion', presented the mental image of a traditional male-dominated manufacturing plant. In comparison, the cable processing plant stood in stark contrast in being a cleaner, quieter and less 'engineering' work environment utilizing predominantly female operators in more labour-intensive activities and tasks. Given these very different operating conditions, job tasks and gender composition, it is perhaps not surprising that there was considerable socio-psychological distance between the two plants, despite the fact that they were physically adjacent. This resulted in a lack of liaison between personnel and poor inter-plant communication. As the manager of the Adelaide operations explained,

> The division between the two buildings was like a gigantic wall. There was no co-operation at all. They were set up with different supervisors. It was their block and no one crossed the line. If you had a job that had to go between the two different areas, it just didn't happen. What we are trying to do is to break all that down so that there is co-operation between the different areas. That is tending to come out with TQM because you are getting different input from different people.
>
> (senior executive interviews, 1991)

Given this historical context, which served to sustain the divisions between the two plants, attention had to be given to improving inter-plant relations before Pirelli Adelaide could consider TQM and the development of more collaborative customer–supplier relations. In an attempt to do this, the manufacturing site manager set about restructuring the existing supervisory hierarchy and developing procedures for encouraging inter-plant liaison. The move towards redefining supervision fitted with the policies developed by corporate management who had prioritized the need for the worldwide development and training of Pirelli supervisors. Prior to these changes, even though all the cables used in the processing plant were produced in the cable plant, the supervisor and leading hands from the processing plant were actively discouraged from entering the other plant. As a result, inter-plant conflicts arose continually, especially over maintenance priorities when critical machine breakdowns occurred (see also Adams, 1990: 31). The conflicts between the plants were further exacerbated by intra-plant rivalries between employees operating on different shifts. For example, there was a tendency to blame previous shift operators for material problems and equipment failures and, in one case, two employees were refusing to talk to one another. Thus, the ability of TQM to improve relations on the shopfloor was constrained by the history and context of intra- and inter-plant rivalry and conflict.

In seeking the development of closer collaborative relationships, the senior executives were also concerned with creating tighter links with their major customers (such as Prospect Electricity, Telecom Australia and Telecom New Zealand) and improving inter-organizational relations. For example, in July 1992 a TQM team had been formed with Telecom Australia and PCAL are hoping to further develop their customer–supplier relations over the next eighteen months.

The formation of teams and the application of quality control techniques

From a local management perspective, the introduction of TQM was viewed as a 'natural flow-on' from the programme which had already been initiated at Minto and Dee Why. The TQM manager and Blakemore Consulting introduced the basic principles of TQM and then, following two one-day introductory training sessions on TQM, three TQM teams were formed at Pirelli Adelaide. Initially, one team was formed from the cable manufacturing plant, the cable processing plant and maintenance. The team from the cable manufacturing was set up to investigate the reasons for the high level of scrap generated by the various manufacturing processes. The team from cable processing was given the objective of collecting and assessing data on the reasons for downtime on their cable cutting and stripping machines. The team from maintenance was formed with the aim of reducing the number of machine breakdowns. It was anticipated that if the time spent by maintenance personnel on breakdowns could be reduced, then it would be possible to allocate more time to preventive maintenance and hence improve machine utilization.

In setting up the initial TQM teams, a facilitator was selected for each team and the relevant production supervisor and shopfloor representatives were asked if they would like to participate. As it turned out, two of the three initial teams had both a supervisor and a shop steward on the team. The teams would meet once a week and the TQM manager (located at the Minto site in NSW) would attend meetings once a month. During the early stages, a number of meetings were allowed to run on an open-ended basis; however, by 1991, clear attempts were being made to ensure that the meetings only lasted for an hour.

Currently, within each team there are a TQM facilitator, whose primary objective is to ensure that things that the team decide need doing get done, and a TQM co-ordinator, who collects all the information, takes minutes, and co-ordinates the meetings. At Pirelli Adelaide there is one TQM co-ordinator – who is also the local training officer – who is on all the teams. This is a full-time position with responsibility for the co-ordination of all TQM activities within the two plants and the development of training (and with award-restructuring, competency-based training manuals). The job involves the collation of all the relevant data collected by the teams and the production of various graphs so that the teams can monitor their own progress. The TQM co-ordinator also provides a training seminar for those new to TQM who are going into TQM groups, and a thirty-minute session on TQM (as part of an induction programme) for new recruits to the company:

> We have a problem with communication, so without having a go at anyone in particular, I put up an overhead and try and explain it. . . . A classic one is a cartoon character with a sword in his hand and all these soldiers coming up the hill. And he is going to fight this huge battle, but there is a bloke behind him with a machine gun who is trying to sell it. What we are trying to say is that if you give us five minutes and listen to people, you will probably turn around and buy the machine gun.
>
> (local management interviews, 1991)

The local management group argued that one of the main aims of TQM was to encourage greater collaboration and teamwork on the shopfloor, and to encourage employees to take responsibility for operations within the plant rather than focus on their individual concerns. In the case of the production manager, the two elements of TQM which were emphasized comprised: first, the resources which can be saved by increasing efficiencies on the shopfloor; second, the way TQM can be used to break down barriers between managers, supervisors and the workforce and between different departments and plants. The view held by local management was that employees involved in TQM generally became the enthusiastic supporters of the programme. However, it was also stressed that decisions on team formation would often be critical factors influencing the success of TQM projects:

> The problems are that you have to be very careful as to who you pick for the teams, the facilitators, and what level you go to. The other thing we learnt was that you have got to be very specific. Concentrate on one area and get data which are very specific to that area so that you can monitor it very closely as opposed to covering a very broad spectrum and getting nowhere. That was the biggest thing that we learnt. The team that was very successful was concentrated on one particular area; one particular problem; itemized; looked at every line; and was very good at concentrating and achieving those goals.
>
> (local management interviews, 1991)

During the first twelve months of operation, the teams which were set up were left largely to their own devices. However, by the middle of 1991 one senior manager commented that the 'ground rules have changed' and that whilst the aim is for the teams to solve the problems, there remained little success in the area of scrap rates and, hence, the manufacturing site manager may be required to get involved. This was because one of the major measures monitored by corporate headquarters in Milan was scrap levels and consequently, considerable pressure was being placed on middle and senior management to show positive improvements in the level of scrap: 'If we don't get some wins up quickly, we could fail and management would have to say that it failed for a reason' (local management interviews, 1991).

The monitoring and evaluation of TQM

In asking the local management group to evaluate the 'success' of TQM at Pirelli Adelaide the following two quotations sum up the general view of the group:

> It hasn't been as successful as it could have been. But it has been a success, even though it has got its knockers. At least people are realizing that things are being done, people are watching it, it is giving avenues to people to have their say. Statistically we can prove it, without adjusting the figures. People aren't falling in line as quickly as we probably would have liked but, underlining everything, it has achieved its goals.
>
> (local management interviews, 1991)

And a more cautious reply:

> I have guarded expectations of TQM because I think it is a principle that can work, that can be effective. I question in some cases our people's ability to make it work. I think that talking honestly, one group has done very well, but the other two haven't achieved a lot in twelve months. I am not directly involved, but I think there might be some group dynamic problems. Like group co-operation. I wouldn't criticize the concept because it is something that we have needed for a long time. It has been difficult to get people to talk about problems in the past because they are too busy. This formalizes the process and creates an environment where they have to. Then you have to change *have to* into *want to*. It is not until you change it to *want to* that you start to get results.
>
> (local management interviews, 1991)

The site manufacturing manager is ultimately responsible for the success of TQM in Adelaide and attends a steering committee meeting held at Minto headquarters every month. At these meetings the executives run through their various TQM teams, highlighting the problems, successes and issues arising, so that there can be some form of cross-fertilization of ideas and critical discussion between the three sites. In this way, it was possible to draw lessons from experiences elsewhere and identify possible ways of improving 'problem' teams. The steering committee is also the body which makes decisions on whether to create, disband or redefine existing teams.

A DISCUSSION ON TOTAL QUALITY MANAGEMENT

Practical lessons on the adoption of Total Quality Management

In the case of Pirelli Adelaide, a number of practical concerns have emerged which can be summarized in the following four general lessons on the management of TQM. First, *TQM is not a universal panacea for problems*. TQM should not be expected to be the panacea for all organizational problems. In the case of Pirelli Adelaide, the history of the two plants and the problems of poor inter-plant communication and liaison were problems which needed to be tackled separately from TQM. Moreover, TQM is often introduced within the context of other ongoing changes and whilst it can be used to assist and complement other changes, it cannot replace the need for other types of change (such as replacing old machinery and/or changing internal reward and recognition systems).

Second, *it is important to continually examine the relationship between technical and social processes in the transition to a Total Quality organization*. For example, restructuring work arrangements through the use of TQM groups may serve to aggravate existing social conflicts on the shopfloor. Whilst participation in TQM is voluntary, operating under systems which have been altered as a direct result of TQM activities is mandatory. Therefore, employees who feel unable to participate because of pre-existing technical or social barriers are unlikely to accept and endorse the decisions made by TQM groups. Such a situation could exacerbate rivalries and conflicts and divide rather than integrate shopfloor employees.

Third, *total employee involvement is rarely practicable at the outset and should be treated as an ongoing objective*. It is rarely feasible to engage all employees at the outset and, therefore, time should be given to the preparation of implementation strategies which may span a number of years. In addition, there may be a number of structural and/or language constraints which may prevent total employee participation. For example, the shiftwork system at Pirelli Adelaide made it difficult to include employees on permanent nights in the TQM teams.

Fourth, *organizations embarking on TQM should recognize the long-term nature of the change and should not over-emphasize the need for immediate quantifiable cost savings*. An over-concern with the costs and costs savings can lead to inappropriate implementation strategies with too much emphasis being placed on short-term financial results. Although there should be some control over the costs associated with the introduction of TQM, conventional cost-accounting techniques should not be used as the sole indicator of the success of programmes which aim at initiating a shift in employee attitudes towards greater involvement and the development of high-trust relationships.

The principles of Total Quality Management in practice

Apart from the practical lessons on the introduction of TQM, the chapter has also highlighted the importance of contextual factors in colouring the nature of the TQM experience. For example, the TQM principle of internal customers and suppliers has only been partially achieved within Pirelli and raises the conceptual problem of how companies should constitute multiple intra-organizational relationships. In cases where an individual or group may have to service multiple customers (such as, next stage in the process as well as opposite numbers on alternate shifts) then a conflict of interest may result and offset the benefits associated with a system based on simple linear customer–supplier progression. In practice, therefore, the assumption that by operating a system of internal customer–supplier relations more co-operative work environments will be created and sustained is a myth rather than organizational reality of operating under TQM.

The establishment of a Total Quality organization has also been constrained by the structural arrangement of the three-shift system. This system has presented a technical barrier to total involvement through preventing a circulation of personnel across the shifts and making it difficult for employees on the night shift to get involved (employees on the afternoon and morning shift generally circulate whilst employees on midnight operations form a permanent shift). Although attempts have been made to involve these employees through organizing early morning meetings these have not proven particularly successful. The night shift operators tend to work in isolation and see themselves as largely independent from management and are generally reluctant to get involved. The TQM principles based on a total management approach to employee involvement and the building of high-trust relationships are therefore not being realized under existing structural arrangements. Although Pirelli recognized the need to improve internal working relationships prior to the introduction of TQM, some rivalry and inter-personal conflicts remain between the permanent night and the other two rotating shifts.

On the question of the shopfloor use of statistical techniques, the empirical evidence suggests that these methods tended to be overemphasized in the training programmes provided by consultant groups. Whilst the groups used brainstorming techniques in their initial meetings, employees did not embrace statistics but, rather, relied on one or two individuals to provide visual graphs to monitor the group's achievements. In addition, employee participation in group problem-solving was found to be limited to those fluent in English language. In this sense, TQM was not well suited to building employee commitment within a workforce that was culturally diverse and demonstrates how TQM has been influenced by the managerial practices associated with the extraordinarily homogeneous society of Japan (Dawson and Palmer, 1993: 134–35). The importance of this issue of cultural relativity is further illustrated by the Pirelli case, where those employees who did not attend the TQM meetings were nevertheless expected to accept the decisions of those that did. In this way, the principle of total 'voluntary' employee involvement was resulting in a form of 'forced compliance' to culturally insensitive group decision-making processes. This mismatch between the concept and practice of TQM suggests that there may be a fundamental flaw behind the organizational assumptions of TQM when implemented within culturally mixed work environments.

With regard to the principle of continuous process improvement the Pirelli case highlights how TQM group-based activity is by its very nature temporary. The notion of continual TQM improvement is a misnomer in so far as the activities are typically project based and oriented to solving discrete problems within a predefined timeframe. Consequently, whilst initial projects may solve major problems and save the company large sums of money, in the longer term the benefits of TQM may become less evident as achievements are restricted to minor system changes. However, this may also detract attention from one of the major aims of TQM, which is about gaining greater organizational flexibility. To have a system which supports the need for continuous change is central to achieving the acceptance of employees for the need to adapt to revisions and transformations in Manufacturing Strategies. In this sense, TQM is about increasing Human Resource adaptability to changing manufacturing arrangements, and creating a link between the technical and social processes evident within organizations. Operating within rapidly changing business market environments, TQM thereby provides a potential method for bridging the traditional divide between technology policy and the strategic development of Human Resource Management.

Outstanding Human Resources issues in a Total Quality organization

The characteristics of TQM and the case analysis of Pirelli Cables both indicate the importance of Human Resource issues in creating quality-oriented companies. At present, however, there remains a general absence of material on the implications of TQM for employment relations, even though the core principles of TQM are marked by an emphasis on employee participation rather than compliance, group performance rather than individual performance, and change rather than stability. For example, W. Edwards Deming (who has been described as the founding father of quality management) emphasized the importance of HRM in his widely cited seven deadly diseases of management and fourteen points for quality improvement (Bowen and Lawler, 1992: 30). Deming has also been highly critical of American managers, arguing that management is responsible for ensuring that employees work smarter and not harder. He advocates that 94% of quality problems are not caused by employee malpractice but by poor management, and that Management by Objectives (MBO) and performance appraisals are extremely demotivational tools which should be discarded in Total Quality organizations (Brocka and Brocka, 1992: 65).

Although the importance of HRM can be identified in the writings of all the major exponents of TQM, senior HRM executives have not typically played a leading role in quality improvement. As illustrated in the Pirelli case, the organizational responsibility for introducing TQM may be assigned to newly formed units and/or TQM co-ordinators. This general lack of senior HRM involvement may also reflect the tendency for a strategic HRM function to be largely absent from modern organizations. As Bowen and Lawler have argued, 'Blocked access denies HR people the opportunity to view the company's overall competitive position. . . . Senior managers may conclude that the human resources issues involved in TQM are too important to be left to the human resources department (Bowen and Lawler, 1992: 32).

However, studies on TQM and the case of Pirelli also indicate the potential for changing existing HRM structures and policies to meet the new requirements of a Total Quality organization. For example, individual performance appraisals are common to the HRM practices of many companies and yet they discourage teamwork and conflict with the philosophy of TQM. Similarly, organizational selection processes are generally based on individual assessments and only pay limited attention to methods for identifying whether potential appointees would function well in team-based work environments. In the case of new recruits, the TQM emphasis on training in problem-solving techniques and group decision-making could form an additional part of the existing HRM training function and thereby promote integration and collaboration between those staff responsible for the quality effort and those concerned with staff development and employment relations. Furthermore, most large organizations have traditionally supported linear career developments – where employees receive hierarchical promotion through a single function – and yet with TQM, the emphasis is on more cross-functional work and the development of a broader range of skills. All these unresolved issues illustrate how many of the existing employment practices are misaligned with the principles and philosophy of TQM. This is turn highlights the need for Human Resource managers to embrace and tackle these issues if they are to play an active and strategic role in the future development of TQM in modern organizations.

REFERENCES

Adams, G. (1990) Redefining the role of the supervisor. Unpublished MBA thesis, University of Adelaide.

Albrecht, K. (1992) *The Only Thing That Matters: Bringing the Power of the Customer into the Centre of your Business*, Harper Business, New York.

Allan, C. (1991) The role of diffusion agents in the transfer of quality management in Australia. Unpublished honours thesis, University of Griffith, Brisbane.

Ballantyne, D. (1992) The new face of service quality management, *Quality Australia*, April, pp. 27–32.

Berry, T. H. (1991) *Managing Total Quality Transformation*, McGraw-Hill, New York.

Berry, L. L. and Parasuraman, A. (1991) *Marketing Services. Competing Through Quality*, Free Press, New York.
Bowen, D. E. and Lawler, E. E. (1992) Total quality-oriented human resources management, *Organizational Dynamics*, Spring, pp. 29–41.
Blakemore, J. (1989) *The Quality Solution*, MASC Publishing, Parramatta.
Brocka, B. and Brocka, M. (1992) *Quality Management: Implementing the Best Ideas of the Masters*, Business One Irwin, Illinios.
Campbell, B. and Davies, L. (1992) *Total Quality in Australia: A Resource Guide*, Australian Institute of Management, Sydney.
Dawson, P. (1994) Quality management in the multi-cultural workplace in Willmott, H. and Wilkinson, A. (eds.) *Critical Perspectives on Quality*, Routledge, London.
Dawson. P. and Palmer, G. (1993) Total quality management in Australian and New Zealand companies: some emerging themes and issues. *International Journal of Employment Studies*, Vol. 1, no. 1, pp. 115–36.
Dawson. P. and Palmer, G. (1994) *Total Quality Management: Breaking The Myth*, Longman Cheshire, Melbourne.
Dawson, P. and Patrickson, M. (1991) Total quality management in the Australian banking industry, *International Journal of Quality and Reliability Management*, Vol. 8, no. 5, pp. 66–76.
Dobyns, L. and Crawford-Mason, C. (1991) *Quality or Else: The Revolution in World Business*, Houghton Mifflin, Boston.
Feigenbaum, A. (1961) *Total Quality Control*, McGraw-Hill, New York.
Fox, R. (1991) *Making Quality Happen: Six Steps to Total Quality Management*, McGraw-Hill, Sydney.
Hames, R. (1991) Managing the process of cultural change, *International Journal of Quality and Reliability Management*, Vol. 8, no. 5, pp. 14–23.
Imai, M. (1986) *Kaizen: The Key to Japan's Competitive Success*, McGraw-Hill, New York.
McConnell, J. (1988) *Safer Than A Known Way: A Quality Approach to Management*, Delaware Books, Dee Why.
Macdonald, J. and Piggott, J. (1990) *Global Quality: The New Management Culture*, Mercury, London.
Oakland, J. (1989) *Total Quality Management*, Heinemann, Oxford.
Palmer, G. and Saunders, I. (1992) Total quality management and human resource management: comparisons and contrasts, *Asia Pacific Journal of Human Resources*, Vol. 30, no. 2, pp. 67–78.
Tenner, A. and DeToro, I. (1992) *Total Quality Management: Three Steps to Continuous Improvement*, Addison-Wesley, Massachusetts.
Walton, M. (1986) *The Deming Management Method*, Dodd, Mead & Company, New York.
Zeithaml, V. A., Parasuraman, A. and Berry, L. L. (1990) *Delivering Quality Service: Balancing Customer Perceptions and Expectations*, Free Press, New York.

7

COMPUTER-INTEGRATED MANUFACTURING: ELEMENTS AND TOTALITY

Malcolm Hill

INTRODUCTION

Computer-Integrated Manufacturing

The application of computer-based systems has become widespread in manufacturing industry as a consequence of their capability to provide a technological basis for improved quality consistency and increased productivity. Continued hardware and software developments, such as Computer-Aided Design (CAD), Computer-Aided Manufacture (CAM), Computer Numerical Control (CNC) and Computer-Aided Production Management (CAPM), have been applied to operate and manage the complete chain of activities from product conception to final product manufacture. In addition to the application of computers to each of these individual areas, Computer-Integrated Manufacturing (CIM) systems have also been developed to obtain benefits from the mutual sharing of design and manufacturing data.

The term 'Computer-Integrated Manufacture' (CIM) has been defined as 'the integrated application of computer-based automation and support systems to manage the total operation of the manufacturing system from product design through the manufacturing process itself, and finally on to distribution; and including production and inventory management, as well as financial resource management' (Harhen and Browne, 1984, quoted in Browne, Harhen and Shivnan, 1988: 33). This definition echoes that of the European Strategic Programme for Research and Development in Information Technology (ESPRIT) some two years before, namely, 'a computer integrated system involving the overall and systematic computerisation of the manufacturing process. Such systems will integrate computer aided design, computer aided manufacture, and computer aided engineering, testing, repair and assembly by means of a common data base' (Commission of the European Communities, 1982, quoted in Browne, Harhen and Shivnan, 1988: 33). The 1984 definition goes beyond the 1982 definition, however, by including computerization of various management tasks in addition to technological functions. Both definitions assume, though, the technological prerequisites of a common database, a data highway, and a series of data exchange interfaces (Brödner, 1990), but the scope of the use of this technology is broader in the 1984 definition.

The term 'Computer-Integrated Manufacture' implies, therefore, a linking together of various computer-based information systems necessary to convert purchased materials, components and sub-assemblies into a range of saleable products. In order to appreciate the objectives and limitations of this integration it is useful to understand the various types of information used in a manufacturing organization; the computer technology that has been developed for the processing of data required for this information; and the various limitations encountered in the development and implementation of these various systems and their integration. These various types of information and systems can be used by a series of specialized departments, and it is consequently important also to have an understanding of the various activities carried out by those departments, and the organizational structure which establishes their relationships. These activities can be summarized as product development, manufacturing, purchasing, stock control, marketing and sales, distribution, and after-sales service.

The next section of this chapter provides a more detailed explanation of these various company activities including the application of computer-based systems and the scope for integration. Particular attention is paid to the application of these systems in the engineering industry, as factories in this industrial sector usually require a wide range of capabilities in the manipulation of both graphical and alpha-numerical data, although some reference is also made to the chemical industry. The third section of the chapter then describes applications of computer-based systems, by means of a number of illustrative real-life case studies, to show the manner and extent to which CIM principles have been used in practice. These case examples provide a chequered picture of CIM utilization, and the main reasons for this take-up are explored in the final section, drawing on information provided in these cases and other published material.

Certain Human Resource factors emerge as critical in the utilization of CIM systems. Most centrally these include labour organization, training to resolve recurring technical problems in hardware and software, and industrial relations. In particular the need to train managerial staffs in the possibilities and practicalities of CIM is revealed as crucial, in order to match manufacturing system design and operation to market conditions, especially when customers' requirements may change during a lengthy period of system implementation.

COMPANY ACTIVITIES

Company activities have been summarized in the previous section as product development, manufacturing, purchasing, stock control, marketing and sales, distribution and after-sales service. From the viewpoint of CIM, however, these activities are often focused around product development, operation and control of manufacturing equipment, pre-production process planning, and production planning and control. This section provides further information on these manufacturing-related activities, the scope for their computerization, and the relevance of other data generated for sales and distribution.

Product development

Product development consists of the creation of completely new products, or the modification of existing items, to meet the changing demands of customers. This task is achieved by the configuration of components and assemblies to achieve specified performance, size, shape and accuracy.

Engineering product development consequently requires two-dimensional engineering drawings which usually consist of the following:

1. a general assembly drawing of the completed product, showing views, sections and details, together with appropriate supplementary information relating to product performance; this drawing usually contains, or has with it, a list of the assemblies, sub-assemblies, components and materials necessary to make the completed product (this accompanying list is usually called the *bill of materials*);
2. a set of drawings for each main assembly and sub-assembly, and the appropriate bills of materials;
3. a set of drawings showing individual components, and the appropriate bills of materials;
4. specifications for materials and purchased commodities.

This design documentation presents information on shape, size and configuration in graphical form; supplemented with alpha-numerical data to define dimensions, and list performance and processing requirements. It provides core information for a manufacturing company, through technical data on component shape, size and accuracy which determine the selection of processes, machines, tooling, materials and operatives. In addition, bills of materials provide the necessary alpha-numeric information on the quantities of particular assemblies, sub-assemblies, components, commodity items and materials required for assembly into the final product. This information, combined with sales requirements, forms the basis for production and purchasing schedules; and when combined with other product design data provides a basis for estimating a significant proportion of product costs.

The application of computer systems to the design process has had a profound effect on the product development function in many companies. Previous designs (particularly of standardized components) can now be stored in computer memories and be rapidly recalled and modified when needed; and various views or elevations of products can be seen instantaneously using three-dimensional software. In addition, the calculations needed to prove or optimize a product's or component's performance or strength can also be carried out using finite element analysis. Moreover, productivity of design personnel has been increased in many industries by the use of Computer-Aided Design and draughting software, particularly in printed circuit board design. Furthermore, three-dimensional models can be created, with the capability to view different aspects of the product's construction.

Computer-Aided Design (CAD) has therefore brought many advantages to product development through the enhancement and acceleration of the design process and CAD systems consequently continue to be diffused throughout manufacturing industry. CAD systems provide the possibilities for further reductions in lead times through the computer-coded transfer of process-related product parameters, which can then provide an interface for the design of tooling and the programming of production machinery and processes. This interfacing of the design and manufacturing functions through the graphical data created at the design stage consequently provides a technological basis for computer integration in design and manufacturing.

Manufacturing equipment

The use of computers in the operation and control of manufacturing equipment has had a major impact over a number of years. In the chemical industry, for example, computers have been used to control various aspects of a process such as the charging (or loading) of predetermined quantities, process temperatures and materials flow. This has been achieved through the development of transducers (devices to convert physical parameters such as temperature, pressure and position into electronic data) and the comparison of process status data with ideal process parameters.

Examples in the engineering industry are to be found in the development of Numerically Controlled (NC) machine tools; but particularly Computer Numerically Controlled (CNC) machine tools where previously prepared programs can be fed into a computer memory adjacent to the machine to locate the cutting tools and workpieces in the correct positions relative to each other. Further application of computerization to this type of machinery provides capabilities for faster programming modification, and proving, either on- or off-machine. In addition, programs can be distributed across a suite of machine tools from a central computer (Direct Numerical Control, DNC). This provides the capability to utilize programming and machining resources more effectively. Modern machining systems consist of Numerically Controlled machining centres equipped with programmable tool magazines and tool changers, which are automatically loaded with workpieces by platen or robot. These are known as 'Flexible Manufacturing Modules'. If two or three modules are linked together to completely machine a defined range of components, they may be referred to as 'Flexible Manufacturing Cells' (FMC). Larger systems using more than three modules with automatic workpiece transfer between them by conveyor or Automated Guided Vehicles, are usually called 'Flexible Manufacturing Systems' (FMS). These are capable of machining defined families of components. Changeover from the machining of one component to another within the family can be very fast, provided that the program has been proved. Flexible Manufacturing Modules, Cells and Systems all have the advantage that throughput time for components is far shorter than if the components were transported between a series of stand-alone CNC machines. These systems can consequently be used to advantage where delivery or cost advantages are to be obtained from the machining of similar sets of components, as outlined in the case of the mining equipment company in the next section of this chapter.

It can be seen, therefore, that the use of computerization in manufacturing equipment has become extremely advanced. This has been achieved through the integration of the mechanical engineering technologies used in machine tool and mechanical handling equipment design, and the electronic technologies and software to co-ordinate the various movements of each of these mechanical components of the manufacturing system.

Process planning

In order that products and components can meet the specifications laid down in design documentation, it is necessary to specify the parameters of each process used to convert raw material to finished product, and the sequences of these processes. This stage of pre-production is frequently referred to as 'process planning'.

In the case of the chemical industry, for example, the process parameters are defined in terms of pressure, temperature and time, with transport provided through pipework and directional control by means of valves. In the case of production engineering, it is necessary to define the machines and tooling configurations to be used at each stage of manufacture together with the directions of the relative paths and speeds to be followed by tool and workpiece. For Numerically Controlled Machinery these relative motions of tool to workpiece at predetermined speeds are defined in a 'part program' for the particular processes to be carried out on a specific component. CAM systems assist in this procedure by storing pre-prepared programs for machining specific geometric features for components of particular size, made from defined materials. These systems also enable the toolpath relative to the workpiece to be simulated on a screen in order to 'prove' the part program and to remove the possibility of clashes between workpiece and tool. In addition, Computer-Aided Process Planning (CAPP) systems may be used to assist in some of these activities, but also to list the process route and associated operations, tooling and transport, to convert purchased materials to finished products.

Since there is a high degree of dependence, however, between the geometric feature of a component to be produced and the relative motions of the toolpath and workpiece to generate that feature, there is a technical possibility of downloading the component geometric data generated at the CAD stage to a CAM system for the generation of part programs. It is this application of computer technology to the design and manufacturing functions, both separately and linked together, which is often meant by the term 'Computer-Integrated Manufacturing'. As explained in the previous subsection, however, it can also relate to integration of the individual items of machining, materials handling and transport; and also to the inclusion of factory scheduling as described in the following section.

Production planning and control

'Production planning and control' is the development of schedules to achieve timely delivery of finished products. It requires a knowledge of each component, of time estimates for each process, and the capacity (usually in hours) of each machine or work centre. This knowledge is essential since schedules list when each component should be processed on each work centre so that the product can be assembled and delivered on time, and meet sales and distribution targets. A production planner must take into account the economic utilization of plant, and make sure that machinery, tools, materials and operators are available to carry out the tasks. Once the schedule has been approved and implemented, progress has to be monitored to see that work is proceeding satisfactorily through the factory, and to enable corrective action to be taken where necessary. To carry out this progressing task, frequent reports are needed from work centres and storage locations in the factory. From these reports the production planner can determine which jobs are being processed on each work centre at any particular time.

Computers are widely used in this activity since they can ease the workload of planning and enable new schedules to be rapidly calculated, and Work in Progress to be reduced. These systems are commonly referred to as Computer-Aided Production Management (CAPM) systems.

The information needed to carry out planning and scheduling tasks usually consists of:

1. A bill of materials, listing the assemblies, components and materials required to produce the end products. This list is usually provided by the design office as part of the 'documentation pack' required to manufacture an item, and it is usually set out so that the relationship between associated materials, components and sub-assemblies can be identified. There is consequently an information link between the design function in the company and the department responsible for production planning and control.
2. A list of the operations required to manufacture each component and sub-assembly listed in the bill of materials, together with the machines, tooling, labour and times that are required. This information is usually supplied by the production engineering or industrial engineering

departments, in the form of processing sheets, operations sheets, or tooling lay-outs.

3. A list of work centres in the factory, and the capacity of each in working hours per week. Shopfloor management usually provides this information.
4. A list of delivery requirements provided by the sales or commercial departments.
5. A list of the components or materials available, or purchasing schedules, supplied by the stock control or purchasing departments, or the stores, within the company.

This information which is mainly alpha-numeric, although graphical information may also be required for descriptive purposes, is used by the production planning and control department to estimate the number of hours required to produce the components, sub-assemblies and finished products, and to allocate jobs to specific work centres. Production planning compares the available capacity at each work centre with the load allocated to each, to establish whether the work allocation is feasible, and also that materials and labour are available. If resources are available, a master schedule can be implemented, and broken down into a series of smaller, more detailed, schedules for each work centre.

Production planning is a very time-consuming process, and a procedure which is very difficult to achieve effectively in practice. Conditions may change radically on the shopfloor as the schedule is being drawn up, and schedules are frequently out of date by the time they are launched. The reasons for this vary, but materials shortages, machine breakdowns and absenteeism are common causes. Because of the frequent need to update production planning schedules, this task is a prime candidate for computerization. Software is available which can generate component requirements from files of bills of materials, and can generate schedules from databases of delivery requirements, workplace capacities, and company priorities and operating procedures. The use of computers for production planning provides up-to-date and realistic schedules, and also the possibility of reducing throughput times and Work in Progress. Production planning programmes can be based on Materials Requirement Planning (MRP), which provides information on the parts and materials required, or the more recent Manufacturing Resource Planning (MRPII) which also takes human and machine resources into account. Further details on MRPII are provided in Chapter 8.

These CAPM systems require data from other company departments, such as parts lists from design documentation, and process sequences and time estimates from production engineering. There is consequently a case for integrating CAPM systems with those of CAD and CAM, in order to achieve a fully Computer-Integrated Manufacturing system, perhaps also integrating with the CAD, CAM and CAPM systems of suppliers and subcontractors where this is technically feasible.

EXAMPLES

Introduction

For the purposes of demonstrating the general level of application of Computer-Integrated Manufacturing, the experience of five mechanical engineering companies will now be reviewed. The five selected are fairly typical in terms of number of employees and product mix. In addition, a recent German survey has suggested that the majority of CAD systems (52%) are likely to be in mechanical engineering, closely followed by electronics (40%) (Lay, 1990). CAD, CAM and CAPM systems had been selected and introduced in these companies during the mid-1980s, and could therefore be considered as being currently implemented and operational. The companies are as follows:

- a designer and manufacturer of aerospace equipment;
- a designer and manufacturer of diesel engines;
- a power engineering company;
- a designer and manufacturer of mining equipment;
- a general engineering company.

The information in each case provides a summary of the systems selected during the mid-1980s, and the reasons for their selection, drawing on relevant cited sources. This material is then appended with information on the current status of the systems, based on discussions between the author and company executives during mid-1992.

A designer and manufacturer of aerospace equipment

This large company designs and manufactures electromechanical and hydromechanical aircraft control systems, which requires the capability to manufacture small batches of components to very high levels of precision. The company is located in the West Midlands, and forms part of an automotive and aerospace engineering group. The systems described in this section of the chapter were originally developed in the early and mid-1980s to meet those manufacturing capabilities and are described in more detail in Ellis and Maxfield (1986).

The company selected a Computervision turnkey CAD/CAM system for design, draughting and process planning, using Computervision software for design and draughting, finite element modelling, NC programming and advanced surface design. Each of these programs is a different application software module, and a process planning/part classification Group Technology software package is also run on one processor. Group Technology provides a basis for the reduction of unnecessary variety in component design, tooling and manufacturing.

Separate CAD workstations are used for conceptual design, schematic lay-outs and general arrangements; finite element modelling and the subsequent display of analysis results; and the creation of component and assembly detail drawings. The data input to the system starts with the creation of a conceptual design and schematic lay-out, both in three dimensions. When these stages have been accepted, general arrangement drawings are developed and, concurrently, finite element stress analysis is carried out on critical items. These results can be displayed graphically, and any required changes can be quickly identified and incorporated. In addition, the methodology is far superior to the expensive non-graphical or two-dimensional methods used previously, which could lead to either over-engineering or subsequent problems in development testing.

The general arrangement drawings are subsequently passed for sub-assembly and component detailing and tolerancing. As workstation availability is limited, drawings are selected according to the potential advantages available from using a computer-aided system, and whether three-dimensional or two-dimensional modelling is required.

The majority of the Computer-Aided Manufacturing activities are related to three main activities, namely tool design in which fixtures to be used on NC machines are designed as three-dimensional models, NC programming, and process planning. The tool design activity requires a wide range of standard parts, and the CAD system allows libraries to be constructed for these parts. In addition, as it is necessary for the tool designer to determine the dimensions of the component at each tooling stage, the fixture models are used by the NC programmers, who use programs for rapid toolpath generation using the component detail generated in graphical format at the CAD stage. The process planning package runs alongside the graphics software, and carries out the rapid generation of process plans and component classification based on Group Technology. This allows families of similar parts to be identified, subsequent standard process plans to be developed, and duplication in components and associated production costs to be reduced.

The trade union organization in the company was involved in various aspects of the system's introduction, including decisions on training procedures and schedules. Potential users, amounting to 130 people, were trained using a one-week introductory audiovisual course followed by small-team training on the system. This was then followed by some three to four days of job-related training when the expected users of the system, namely draughtsmen, designers and NC programmers, were subsequently given the opportunity to reproduce an existing working drawing and the production of new drawings. The company's experience suggested that the users gradually achieved proficiency over six months provided that opportunity existed to spend at least 50% of their time on the system.

The company has found that the use of the system has led to significant improvements in design productivity. Finite element modelling (a technique used in component stress analysis) is now faster by ratios varying between 10:1 and 50:1, and NC programming is faster by about 5:1, compared with previous methods. For certain types of draughting amenable to parametric programming, time savings of about 5:1 have been obtained, with further savings anticipated through retrieval of similar parts. The requisite database of some 5,000 components was expected to take some four years to compile.

The company did not follow conventional methods of financial justification, as many of the savings were viewed as intangible or fundamental to continued operation in advanced technology, although all benefits were recorded wherever possible. Since the size of the installation is based on steady-state requirements, peak workloads have required the operation of a double shift to increase utilization. No industrial relations problems have occurred, probably because of the voluntary basis of this operation.

Since the introduction of this system in the mid-1980s, the company has continued to extend the applications described above, and to take advantage of computer-aided engineering and CIM developments, particularly in the areas of solid modelling and integrating with other systems such as DNC, Tool Management, inventory control and Cellular Manufacturing scheduling. To maintain competitive advantage, it has also become necessary to increase the number of support staff and up-rate the basic system. Catia has replaced Computervision as the main CAD/CAM system.

A designer and manufacturer of diesel engines

This company is a large design and manufacturing establishment, located in the East Midlands, which is part of a larger automotive group. The company has to produce a wide variety of end products to meet market requirements, but also has to produce many common components in large volumes in order to remain price competitive. The company has consequently used microcomputers extensively in its machine and assembly shops, in order to monitor the tooling and working status of its high-volume production. The variety of components to be tracked is also very high, requiring the use of minicomputers to provide the necessary parts list and assembly data, and to monitor progress, from a schedule developed from a logistics and planning system stored and processed on a mainframe computer.

The design function is a core factor influencing the company's present and future success in the marketplace, both in terms of product quality and the speed of new product introduction. It has consequently been necessary to choose a CAD system with care, and most of the remainder of this description is a summary of the selected system described in Reader (1986). The company selected a twelve-workstation Computervision system with appropriate design and draughting software. The workstations are used primarily for work on new engine projects, with existing drawings loaded on to the database, where necessary. The situation for manufacturing aspects has changed in the last few years, however. Three additional workstations were used for manufacturing applications, particularly for jig and tool design. This work was usually carried out in parallel with a design project, making use of the existing component geometry in the engineering database.

The other set of computer applications in manufacturing occurred in Numerical Control programming, where the type of application was very often different from that of design. This was because Numerical Control provided a capability for manufacturing flexibility across a wide range of components, with advantages possible for components sharing similarities in design features. As new machines were purchased, therefore, it was usually the case that data were required for a range of similar components, rather than specific components for particular design projects. However, the emphasis in recent years has been to outsource much of the machining operation, and consequently CAD has become more concentrated on the design and development process and less on machining.

A power engineering company

This large company, located in the East Midlands, is a leader in the design, development, manufacture, supply, erection and commissioning of major steam generating equipment. The company has been part of a major power engineering group for several years, which has recently merged with a leading aero-engine manufacturer. Principally, the company markets large boiler plant and associated equipment for power stations worldwide. Its environment is essentially one of heavy engineering, manufacturing and assembling components from 'hand-size' to 'house-size'. The materials, handling equipment, machine tools, and heat treatment equipment are mostly large and heavy requiring extensive working space and buildings.

There has been a decline in the orders for power station plant in recent years both in the UK and abroad. At the same time, customers have been calling for shorter time-scales with deliveries being strictly on time and within cost. They are also demanding evidence of first-class project management, and production control systems within their suppliers, to give confidence that projects will be completed as required. The company has long recognized these needs and has for some years invested heavily in plant, lay-out, and the adoption of a companywide IT strategy to improve its planning, production control and scheduling. It also participated in a Teaching Company Programme with its local university to help with parts of this implementation, and a description of that strategy from which this case is summarized is provided in Suter and Leary (1989).

The company has implemented a project management resource/ programme package ('Trackstar'), on a Prime mainframe computer, which had previously been installed to drive the company's CAD system. Trackstar serves both the high-level projects department (which liaises with customers), and also the planning and product control department, one level lower, which plans and controls workflow in the production shops. In this make-to-order industry with project time-scales of several years or more, it is essential that the period is split into smaller, key stages with key dates to monitor. The planning and product control department uses Trackstar to express the manufacturing stages of the contracts as co-ordinated plans for each workshop showing the key dates they must meet. To be more useful at shop level, however, these broad plans must be further defined in terms of operations sequences based on available resources (capacity planning/resource scheduling). Appropriate systems for this scheduling are now well advanced in a number of shops but the next sections of this paper focus on two of the main manufacturing shops that differ in product, process and control needs, namely the machine shop and the heavy fabrication shop.

The machine shop

This large shop has a skilled workforce and a wide variety of machine types and sizes. It serves as a vital 'feeder' to the other ('product') shops, providing parts from hand-size to those of several tons. Batch quantities and lead times vary but these times are relatively shorter than in the heavy fabrication shop. The operations are usually sequential. The machine shop must regularly examine its current and forecast loading on its key resources (work centres) if it is to feed back overloads and possible delays to the key target dates from the planning and product control department. Programme changes do occur and again it is vital that the consequences can be readily assessed and any necessary rescheduling be co-ordinated. The shop therefore needed to have a system which would hold resource capacity details for any period, would include appropriate prioritizing rules, and have good reporting facilities to allow feedback.

It was also important to know clearly which jobs should be completed in the following week (or required period), and in which priority order. These 'Work- to Lists' (WTL) would co-ordinate with the issue of job tickets. It was also desirable that Work in Progress (WIP) information available from an existing ICL mainframe computer could be used to show which WTL operations were actually issued with materials being available.

It was realized that as many of the shops had quite different operational characteristics and scheduling needs, each shop should operate its own scheduling software that met its particular needs. However, the software was also to be capable of exchanging data with the mainframes to get WIP information and the Trackstar planned target dates. To effect this, the company installed a Local Area Network (LAN) so that each shop could be connected with microcomputers, and receive the necessary data via the links with the mainframes and schedule its own situation.

For the machine shop, more than ten software packages were considered. 'MICROSS' from Kewill Systems emerged as the favourite in view of its abilities for finite capacity scheduling, load reporting, time analysis, residual and autoscheduling, multi-calendars, job status, overdue jobs and the capability to produce the required Work-to Lists. MICROSS was also available for extensive evaluation and proved to be relatively inexpensive at about £2–3 thousand.

The heavy fabrication shop

The heavy fabrication shop is also very large, being able to fabricate pressure vessels of up to 200 tonnes. In this shop, very large, thick steel plate is hot rolled into cylindrical forms to be welded together and joined with other formed and fabricated parts, to build into pressure vessels.

Unlike the machine shop, this 'product' shop is very labour intensive (having about sixty people, many being skilled welders). It handles very large parts, and has much longer lead times, often into months. It has considerable concurrent activities occurring on a number of vessels and demands greater flexibility in the sequence of operations. The precise work content of any job is also more difficult to quantify.

The planning and scheduling was previously based around bar charts produced manually for each vessel. These were time consuming, difficult to co-ordinate and did not reflect the resource requirements of the shop. As it is important that this shop's schedules match with the key dates from the planning and product control departments, software offering 'what-if' analysis, resource loading and bar chart representation, and networking logic was therefore thought most appropriate. A PC-based, project management type package, 'Pertmaster Advance', was initially selected but later found to be limited and unable to handle all the data required. The knowledge gained with MICROSS meant this was reviewed and chosen, although not ideal, as a workable, acceptable alternative.

Various investigations clearly showed that the operating characteristics and requirements of each shop varied considerably. Adopting a LAN approach at shop level for scheduling and linking this with the mainframes has created a flexible hardware configuration. It has also given each shop the potential to use different software unique to its scheduling needs so long as it can use the WIP and key dates data from the mainframes. MICROSS has been used in this flexible approach to good effect. The use of this structure has consequently created the means of integration between a mainframe-driven master scheduling system and Local Area Networks related to the specific conditions in each manufacturing shop.

This case reveals, therefore, the role played by important customers in influencing innovation (in this case the adoption of a computerized production planning and control system), and the scope for selection of high-level mainframe software serving different departments with consequent possibilities for cost sharing. In addition, the case demonstrates that the specific operational needs of production planning and control can be carried out by networked microcomputers having access to data from high-level software. The preferred microcomputer software in terms of output may lack the capability to process large quantities of complex data, however, leading organizations to make a choice of software based on minimum inconvenience rather than maximum advantage.

A designer and manufacturer of mining equipment

This large company, located in the west of Scotland, is a manufacturer of coal-cutting machinery for both domestic and overseas customers. The components used in the build-up of these machines chiefly include prismatic items such as gearbox casings, beam-shaped components such as booms, and rotational components such as shafts and gearwheels. In addition, the company's products also require electrical and hydraulic power and control units, built to the demanding safety standards required by the mining industry. The company's experiences in Advanced Manufacturing Technology have been described previously in Hill (1985), Hill and Woodward (1987: 78–81), and Boer (1991: 133–9), and the information provided in this chapter draws extensively from that published material.

In 1984 the company's stock of machine tools at its main site was 340 with a further 100 machines located in feeder factories. Within this total stock the company used 40 boring machines but also frequently required feeder and sub-contracting capacity of up to another 40 boring machines, as the load on its resources fluctuated. Sub-contracting brought problems associated with control over quality and delivery requirements whilst additional transport costs were associated with the use of feeder factories. These factors influenced investment in a Flexible Manufacturing System as a means of increasing the company's in-house boring machine capacity in a manner consistent with high labour productivity.

The degree of sophistication of the company's machine stock had also been changing significantly over the previous twenty years. Prior to this, the company had used stand-alone machines operated by highly skilled employees, but during the mid-1960s Numerically Controlled point-to-point drilling machines were purchased to reduce the company's investment in drilling fixtures. In addition, a small Numerical Control department was set up in the company, but no attempt was made to move into other areas of Numerically Controlled machining because of difficulties likely to be encountered in program editing.

In the 1970s the company purchased CNC turning machines and also established a 'machining cell' for the manufacture of certain steel components used to build gearbox casings, shells and housings, since these components were particularly difficult to manufacture under automated conditions in view of their weight and variation in material hardness. Care was also taken to select machines capable of providing high labour productivity, since it was proving difficult to recruit and retain skilled boring machine operators at that time.

As a consequence of twenty years' experience in the use of Numerically Controlled machinery, assimilated in a series of stages, the company possessed the necessary corporate technological expertise to explore the potential of Flexible Manufacturing Systems. The need for this technological expertise became apparent as a result of changing market requirements for shorter delivery lead times, and fast rates of product innovation.

Export markets have recently accounted for 40% of the company's sales, compared with 20% ten years previously, and this success in exporting has been dependent upon the ability to quote a delivery time of some five months. However, total throughput times for the company's products were of the order of ten months. These demands had conventionally been met by planning provisioning of delivery-critical components in line with forecast orders, with recent improvements being achieved by a computerized MRP and capacity planning system. It was apparent, however, that any method which could reduce production lead times would be welcomed by the company, particularly for the manufacture of shells and castings which took some six months to machine and accounted for an appreciable portion of the product's cost.

As well as customers requiring shorter delivery times, it also became apparent that the market was demanding a faster rate of product innovation. In the decade between 1972 and 1982 the company had introduced two new ranges of machines, whilst during 1983 alone the company had launched three new machine ranges. In addition to these market requirements for reduced lead times and the capability to change over to new products, shorter production time capability was also attractive as a means to reduce Work in Progress inventories and associated holding costs.

As a consequence of these market and product costs savings possibilities, an initial study was carried out on fourteen of the company's thirty different types of shell castings. A more detailed study was carried out on seven of these fourteen castings with a longer product life. Initial comparisons were between stand-alone conventional machines, stand-alone CNC machines, and a Flexible Manufacturing System. Stand-alone conventional machines were initially rejected because of their low levels of labour productivity and stand-alone CNC machines were subsequently rejected because between 200 and 300 tools would be required per gearbox, with associated difficulties in Tool Management. A final factor favouring the selection of an FMS was its capability to schedule sets of different components through the system to meet assembly requirements.

The system consists of five CNC machining centres and one CNC facing head machine. The workpieces are automatically loaded and unloaded on to the machining stations, and transported between stations by means of a shuttle car. The system is served by a workpiece loading department which loads the castings into fixtures prior to the first operation and also rearranges the part-machined components in the fixtures as necessary. In addition, the system has its own tool-servicing and tool-monitoring facility.

It was intended that the system would reduce the total production lead time for the machining of castings from some six months to two months. This reduction in lead time was consequently expected to lead to a Work in Progress reduction to a third of its previous level of 400 castings. In addition, the castings were to be scheduled through the system in sets to meet assembly requirements, rather than scheduled to maximize machine cutting efficiency. The system took some two years to build and eighteen months to install, with some of those activities running in parallel over a twelve-month period.

The total investment cost for the FMS including machining stations, materials handling equipment, tooling, fixtures, computers and software was in excess of £6 million. Of this initial investment 22% was covered by a Regional Development Grant due to the company's location in the west of Scotland, and 25% of the 78% balance was covered by Selective Financial

Assistance under Section 5 of the Science and Technology Act. In addition, part of the company's training costs were met by an EC grant designated for companies located in coal or steel communities. The system was found to be economic compared with stand-alone conventional and CNC machines in view of an expected five-to-one reduction in production lead times compared with stand-alone CNC machines, thereby providing lower total production and inventory holding costs.

The company did encounter a number of technical problems during the implementation of the system, however, related to shortcomings in one of the machining stations requiring a major redesign of that unit; and some communication shortcomings between the tool-presetting station and the host computer. In addition, assembly and proving of many of the programs caused severe problems, and the NC part programs had to be tested, proved and optimized at the system. At the same time, however, there was a significant reduction in the company's order book, so expected sub-contract labour was consequently not required. Anticipated reductions in machine times and average lead times were also achieved, together with anticipated reductions in manning levels, consistent with the agreement with the trade unions that there should be no redundancies attributable to the system. Problems associated with programming did act as a bottleneck in reprogramming for new components; and market conditions also changed, which placed an extra demand on the programming section for the production of new components.

In recent months, the market for mining equipment has continued to decline to such a level that the use of a stand-alone machining centre with a reliable facing head is now more cost-effective than the FMS system for the number of units required. The company has consequently decided to transfer production to stand-alone CNC machines and sell the FMS system. There have also been several other factors of a strategic nature which have influenced the company's decision. These have included an increase in international manufacturing collaboration agreements related to increases in export sales, higher volumes of component manufacture being transferred to the company's overseas subsidiaries, and a general trend to sub-contract more machining. In addition, the technical problem on the FMS facing head was never fully resolved, whilst a decision to merge two of the company's factories provided access to reliable facing capacity.

A general engineering company

The company produces a diverse range of capital equipment to customer order for the metal processing industries (for example, slitting mills and rolling mill conversions). Investment in Advanced Manufacturing Technology has been undertaken in two areas, namely CNC machining, and production planning and control; and the company is currently considering investment in Computer-Aided Design.

The machine tools traditionally used by this company were manually set and operated by skilled craftsmen. This was cost-effective for much of the one-off and small-batch production work carried out by the company, but for medium-batch and identical or similar work repeated at intervals of time (perhaps weeks or months), Numerical Control offered a better alternative.

As a result of the potential benefits offered by Numerical Control, the company decided to purchase a CNC lathe and milling machine. The former has been used for the production of rotational components and the latter for the generation of flat surfaces, recesses and bores. CNC was selected because it provided the capability of shopfloor programming for new components and editing of previous programs used for similar products. This capability was particularly attractive for this small company as it did not wish to incur the additional overhead cost of a separate Numerical Control tape preparation department, and also because it wished to capitalize on the traditional skills and technical expertise of its machinists.

The company's machinists took readily to the new technology, and were capable of programming the machines after a three-week training course provided by the vendors. The company was able to acquire one of the machines at a very competitive price, and was also able to benefit from a DTI-supported programme to aid engineering companies investing in Advanced Manufacturing Technology. This aid enabled the company to receive a grant which covered up to 33% of the initial costs of the machine.

It is important to note that the hourly rate for the use of the new machines was far higher than for the older plant. This was a result of higher depreciation costs arising from the more expensive outlay for the purchase of the new machines, and additional overhead costs from the increased resources required. However, because of the shorter set-up times on these machines once the program had been proven, and the more consistent cycle times throughout the working shift, the cost per piece was lower for appropriately selected components and batch sizes. Additional CNC machines have been purchased in line with company expansion, following the initial investment.

A further area for computer-assisted improvement was the production planning and control system. Planning work had previously been carried out manually with priorities for each work centre being estimated from product delivery dates. This task became increasingly complex and time consuming as company turnover increased, and sometimes became arbitrary in view of the time pressures to issue a plan. As jobs progressed through the factory it was necessary to collect information from each work centre to set new priorities. This was usually done on a weekly basis to enable the information to be tabled at a weekly review meeting between the production manager and his foremen. At this meeting, new priority lists were set for each work centre for the forthcoming week. Because of the time required to gather all of this data manually, the information could often be two days out of date when the meeting commenced. In addition, information gathering and the associated calculations necessary for the stores and purchasing office were carried out manually, which sometimes delayed the presentation of status reports on inventories and purchasing schedules.

In order to obtain more consistent planning, therefore, and also more accurate progressing information, the company decided to investigate the possibilities offered by computer-assisted production planning and control. The objectives were to improve delivery capability by enabling schedules for each job to be translated into Work-to Lists for each work centre, ranked on a priority basis according to delivery requirements. In addition, the faster input of job progress information into the system would enable new priorities to be established in a shorter time. Furthermore, the company was hoping for increased sales as the national economic recession became less severe, and it wanted to be able to meet these sales requirements without a lengthening in delivery dates or an upturn in costs.

It became apparent, however, that it was extremely difficult to quantify many of these anticipated benefits arising from improved information, and to express them in terms of financial advantages with which to justify the costs of purchasing and implementing a computer system.

It was also expected that the use of a computer system would improve the productivity of production control staff, enabling extra turnover to be met without increased staff. Furthermore, it was expected that Work in Progress as a proportion of turnover would be decreased as a consequence of improved production control. The reduced financing charges for Work in Progress could consequently be used as a further financial justification for the system. Finally, it was intended that stock control and purchasing operations would be significantly improved.

For this company, it was also particularly important that the software package related to a 'jobbing' environment, with manufacture to customer order, rather than to maintain a stock of fairly standard products but with perhaps different features or dimensions for each customer. Thus a job library was considered to be useful to hold standard product details which could be called up and modified when required.

The software had to assist with the following tasks:

- order analysis and processing (creating and maintaining all of the engineering information generated by the manufacturing and design engineers);
- purchasing control (including the production of purchase order documentation);
- stock control (including parts lists and possibly bills of material);
- production planning and control, encompassing: Work in Progress control, loading and scheduling, operations routeing, and factory paperwork;
- job costing and analysis.

The company carried out an initial review of fourteen systems, matching their capabilities against the firm's detailed requirements in purchasing, stock control and production. These fourteen systems were then narrowed to four. Detailed discussions were carried out and visits made to user sites in the engineering industry. Particular attention was paid to the hardware and software capability of the vendor, and the maintenance and support required; the track record of the vendor on the user site, particularly in the area of support provided during implementation and training; and the business survival capability of the vendor. From this screening, two systems offered by the same vendor were subjected to financial analysis based on the following cost headings:

- hardware purchase;
- software purchase;
- hardware maintenance and after-sales service;
- software maintenance and after-sales service;
- training (time and cost);
- installation and implementation (time and cost);
- consumables;
- users (in purchasing, stores, drawing office and production office);
- Work in Progress and stock savings;
- personnel savings;
- grants.

These data were then subjected to a discounted cash-flow analysis over an estimated system lifetime of eight years.

A system was finally selected which could convert job network plans into priority Work-to Lists for each work centre, and could rapidly calculate new priorities. Implementation took longer than expected and more supplier support was required. The problems of implementation were finally solved by the recruitment of a systems engineer to be responsible for the project.

The company has found that its control over the production function has been radically improved. Operator and supplier performance can be rapidly assessed, information on stocks and Work in Progress is more readily accessible, and up-to-date lists on job lateness are now available. Furthermore, the information available allows on-line job costing to be carried out, and variances with estimates to be quickly identified. Finally, in view of the ability to rapidly monitor progress in both purchasing and manufacture, clerical resources have been more effectively used in chasing suppliers rather than routine administration.

The company has found it difficult, however, to achieve its target for reduced Work in Progress, and is sceptical of some vendors' claims that Computer-Aided Production Management systems can be easily paid for through reduced Work in Progress. The company sees the major benefits from Computer-Aided Production Management as being

- better workflow control;
- better shop loading;
- vastly improved access to management control information in all operational areas.

The company's management point out that although these advantages can be easily identified, it is very difficult to quantify them in financial terms. However, the increase in operating efficiency has led to the recent decision by the company to invest in a second-generation system using the latest releases of the same software and utilizing a linked network of PCs which can now accommodate up to 100 users. Sales order processing is being introduced and direct links to accounting packages are planned.

DISCUSSION

The cases described in the previous section of this chapter illustrate that the major developments in application have apparently been directed towards improvements in each of the specific areas of design, manufacture and production management, rather than integration between them, although some integration has taken place between design and manufacture. The reasons for this, together with the appropriate management issues to be addressed fall into the following areas:

- technology;
- markets;
- government policy;
- financial justification;
- industrial relations;
- training and organizational arrangements.

Each of these topics is discussed below, drawing on material available from the literature, and supplementing this with information from the case studies described above. The general impression derived from these case studies is, however, contrary to some of the literature. The case material suggests that progress towards integration in computerized manufacturing is shaped more by market pressures and technological constraints than industrial relations, training or organizational issues. Human Resources of the requisite quantity and quality are clearly required to solve these technical and market problems, however.

Technology

The hardware and software systems that have been developed and implemented have had the improvement of their own particular field of operation as their major objective. Computer-Aided Design packages have been directed towards enlarging their capabilities in the design function through faster calculations and more refined heuristics, whilst associated computer-aided draughting systems have been directed towards improved speed, convenience and clarity of presentation during the draughting process. Clearly, CAD packages have some application in manufacturing, as mentioned in two of the case studies, particularly in terms of tool design; and the direct linking of product design to manufacturing includes the presentation of product graphical data for tool design, CNC part programs and process plans. In a recent West German survey of CAD systems it was found that about 20% of all CAD systems installed had CAD integrated with NC programming capabilities (Lay, 1990). A higher level of further integration has probably been prevented by the complexity of the process planning function, where attention is devoted to the compilation of libraries of data relating to tool and material behaviour under manufacturing conditions, and the installation of CAD systems in organizations engaged only in design.

Technical difficulties encountered in integration were apparent during the introduction of the Flexible Manufacturing System in the mining equipment company mentioned in this chapter, where it was necessary to integrate cutting-tool management, fixture management, machine operation, and materials handling sub-systems. Many of the problems which occurred were due to technical problems in the manufacturing hardware, including the areas of interface between mechanical drive units and control systems, and software errors in each of the sub-systems and in the integration between them; and these problems were found to be typical for many of the early FMS systems implemented in the UK (Hill, 1985). The results of subsequent research in western Europe indicate that technical problems still remain during the implementation of FMS systems (Boer, 1991).

The information obtained from the five companies described in this chapter also suggests that the pace of integration between manufacturing technology and production planning and control systems has been comparatively slow, probably because both sets of sub-systems tend to deal with different types of data: manufacturing technology systems are concerned with processing data such as machine spindle speeds and tool design, whereas production planning and control is concerned with workflow and timetabling. Although there are areas of interface between these two types of data in terms of materials lists (30% of German CAD systems generate bills of materials (Lay, 1990)), and process routes and processing times, and many types of scheduling system are now graphical in format, manufacturing technology systems are still more concerned with the presentation and manipulation of graphical data than are their counterparts in scheduling and control. According to a recent German survey, only 1% of CAD systems (which predominantly use graphical data) can be linked to a production planning and control system (Lay, 1990).

At present, therefore, it appears to be the common perception that more benefits are to be obtained in the improvement of each sub-system rather than in sub-system integration. This is discussed in more detail in a section below covering financial justification, but the information available from the case studies described above appears to support the view expressed by Bullinger (1990) that there is still a need for more proven and standardized CIM concepts available for use in the marketplace, as a prerequisite for more widespread integration.

Markets

The perceptions and realities of market conditions have had a profound effect on the development, implementation and success of individual computer applications in manufacturing and their linking together into an integrated system. Computer Numerical Control, for example, is now an established manufacturing technology, enabling companies to meet market requirements for consistent quality of production once part programs have been proven. Prices can become more competitive because of cost reduction through better product consistency, reduced changeover times, and faster processing speeds available through the improved rigidity of the more recent designs of machines. Furthermore, shopfloor personnel have been willing to learn the relevant programming techniques, and their representative trade unions have supported the diffusion of this Advanced Manufacturing Technology and have not adopted a stance of opposition.

More competitive delivery times have also been possible through the shorter processing times available from CNC machines and the potential for combining several processes on to fewer machines, with consequent reduced travel time. The most significant reduction in production lead times, however, has been achievable through the integration of CNC machine tools into Flexible Manufacturing Systems, and the consequent reduction in transport and queuing times at each machining centre. The effect of this has been to reduce 'door to door' times, rather than only 'floor to floor' times as in the case of CNC, and this reduction has also provided possibilities for reduced Work in Progress and stockholding, which can also lead to lower costs (and consequent prices) for the same level of customer service. These market advantages, therefore, can be viewed as directly attributable to integration of the proven benefits of CNC, but within the carefully defined boundaries of machining technologies and component envelopes to match focused manufacturing technology to specific product ranges.

In the case of the mining equipment company, however, it was found that the time required to commission the FMS was far greater than the time required to commission an equivalent quantity of stand-alone CNC machine tools, and this extended implementation time was directly attributable to the technical problems of integration. In addition, market conditions significantly changed during and after the time of implementation, and the FMS was not the best manufacturing system to suit the new market conditions. This apparent mismatch between the originally specified FMS and changing market conditions has also been apparent amongst other users of these advanced types of integrated manufacturing systems (Boer, 1991: 222–4).

CAD systems have been introduced to improve the productivity of engineering designers and consequently reduce design costs and their impact on product prices, particularly where design skills, techniques and experience continue to be a scarce commodity. Improvements in design productivity have led to satisfactory returns on investment for successful users, but initial investment costs were frequently very high although unit prices for CAD systems are now decreasing. Of more importance, however, has been the capability of the CAD systems to shorten the design lead times of new products, and consequently also increase the pace of innovation in the product mix (Ayres, 1990: 152–3). However, the implementation of some systems occupied an extensive time period as a consequence of systems problems, inadequate specification and benchmarking, and some industrial relations problems.

Similar market objectives were sometimes laid down for CAM systems, namely to reduce pre-production planning times. These savings have sometimes been difficult to establish in practice because of the influence of other factors such as changes in demand and capacity. In several cases, the anticipated savings in pre-production time using CAM systems have been significant (Ayres, 1990: 152), and also achieved as a direct consequence of the integrated use of data generated at the design stage, as in the aerospace equipment company described in this chapter. In the case of the designer and manufacturer of diesel engines, however, market conditions and decisions on Manufacturing Strategy were causing the company to withdraw from component manufacture and thereby become more interested in developing the CAD system alone, rather than further integration between CAD and CAM.

It was also expected that CAPM systems could provide the requisite scheduling data for shorter and more consistent delivery times, together with reduced Work in Progress and consequent cost, although the experiences of the small engineering company described in this chapter suggest that improved control over workflow is the major advantage to be gained from the use of these systems. In the case of the power engineering company, improvements in workflow and delivery reliability were contingent upon the development of a LAN to link production planning and control information between the various manufacturing shops, and the selection of a software package which could be shared by both the manufacturing and the projects departments.

Government policy

Government policies for industrial financial support have exerted a significant influence on the selection and financial justification of computer-aided Advanced Manufacturing Technology, particularly when combined with financial support for investment and training in different geographical regions. In a study of Flexible Manufacturing Systems installed in the UK in the mid-1980s, for example, it was found that a major factor influencing the profitability of the investment was a combination of two types of government-aided financial assistance, the first related to support for Advanced Manufacturing Technology as such (approximately 25% of the investment) and the second related to investment in regions of high unemployment (a further 25%) (Hill, 1985). The mining equipment company described in this chapter was able to obtain such financial assistance which made the investment in FMS even more attractive.

In addition, in the case of the small engineering company investing in Computer Numerical Control for the first time, the combination of government financial support for investment in Advanced Manufacturing

Technology combined with competitive prices for CNC equipment, caused the investment to appear as an attractive proposition. As Advanced Manufacturing Technology has become more widely accepted, the necessity of government support has perhaps been seen as less important for its diffusion, although this perception is open to question, particularly in conditions of industrial recession.

Financial justification

The financial justification for computer-aided Advanced Manufacturing Technology has normally been achieved in terms of reduced labour costs through the increased productivity of the more advanced equipment, and reduced WIP costs as a consequence of shorter lead times. These savings were then annually discounted over a defined time interval to then establish whether a 'hurdle' rate for return on investment had been cleared.

In many companies, however, it is frequently found that many of the benefits ascribed to Advanced Manufacturing Technology cannot be easily financially quantified, although the investment costs are only too apparent and usually very high (Bullinger, 1990). Many of these benefits relate to capabilities which could not be achieved previously, such as more consistent clarity of engineering documentation and wider heuristic capabilities in engineering calculations. It is sometimes possible to quantify these improvements through expected increased shares in new markets and increased sales.

Many of the benefits available from Advanced Manufacturing Technology, which also frequently cannot be financially quantified, are also of fundamental strategic importance to the future of the company, and the effects of this intangibility on company strategy have been discussed at length by Kaplan (1984, 1988) and Johnson and Kaplan (1987). As markets require shorter delivery times, for example, the use of CAD systems enables pre-production design times to be reduced, and the use of FMS provides a technological basis for the reduction of throughput times. If market conditions require these changes, therefore, one method of financially justifying the relevant investments may be through estimating the financial consequences through lost market opportunities if the changes are not implemented. In most of the cases described in this chapter, financial justification of Advanced Manufacturing Technology was found to be difficult, although detailed attempts were made to justify the FMS in the mining equipment company, in terms of productivity savings and reduced WIP. In the case of the aerospace equipment and power engineering companies, investments in computerized systems were seen as central to the company's strategy, and customer driven by one important customer in the latter case. The expected benefits of improved workflow from CAPM systems were also obvious in the power engineering and small engineering companies, but difficult to quantify in terms of reduced WIP, as this parameter is dependent upon a wide range of factors such as demand against capacity.

Industrial relations

The implementation of new technology is frequently associated in the public mind with a high degree of industrial unrest, partly as a consequence of the disturbances in the newspaper industry during a time of rapid technological change during the 1980s. In manufacturing industry, however, shopfloor resistance to the introduction of new manufacturing technology has not been particularly strong, particularly in the area of CNC. In the case of the small engineering company the shopfloor personnel were apparently enthusiastic in their response to learning NC programming techniques, and the workforce in the mining equipment company were willing to work on the FMS provided that redundancies did not result from the implementation of this equipment.

This lack of resistance is partly due to the workforces' enthusiasm to acquire new marketable skills, and partly due to 'new realism' in trade unions, as market conditions and industrial relations legislation have reduced the opportunities for, or possible gains from, industrial conflict. In addition, many trade union officials actively support the implementation of Advanced Manufacturing Technology, as a way for their members' employers to remain viable and therefore for jobs to be retained, and they welcome the scope for enhancing their members' knowledge and skills and consequent employability. These options have become particularly attractive when agreements have been made that the implementation of the Advanced Manufacturing Technology would not lead to job losses, and several employers and trade unions have signed new technology agreements specifying various facets of objectives and implementation.

According to a trade union official interviewed by the author, some resistance had been encountered in the implementation of Computer-Aided Design systems, however, as a consequence of the radical differences in working practices expected from designers and draughtsmen in a CAD environment. These changes in working practice have arisen mainly as a consequence of the high cost of CAD equipment and the necessity of using this equipment more intensively on a shiftwork basis to justify this investment. This change from daywork to shiftwork, in a white-collar working environment, has consequently led to some employee resistance. In addition, this resistance has increased in some cases as a consequence of limited opportunities for training, because of the small number of available workstations, and the necessity of maintaining an existing work schedule. In the present economic and employment climate, though, designers and draughtsmen were prepared to operate the equipment before conditions of employment had been finalized and agreed between their employers and their trade union representatives.

In the case of the aerospace engineering company described in the case study section of this chapter, however, it was found that the design staff adopted a co-operative stance, probably because participation was encouraged in equipment selection, and shiftworking was voluntary.

Training and organization

The implementation of Advanced Manufacturing Technology requires a high level of relevant training in both hardware and software in order to acquire the necessary skills in successful system operation, and the solution of technical problems which may arise. Many of the requisite training programmes are provided by equipment manufacturers, but implementing companies are required to timetable training schedules in order that expertise can be gained consistent with a satisfactory level of equipment utilization and a minimum of disturbance to existing schedules of output. This may require the recruitment of additional personnel, as in the case of the small engineering company, or the supply of temporary staff through such arrangements as a Teaching Company Programme, as in the case of the power engineering company.

What appears to be of equal importance, however, is the training of senior managerial staff within the company in the skills of strategic analysis and the process of implementation, to ensure that systems are selected to support the company's present and future competitive advantage in terms of quality, delivery or costs. This selection, in its turn, may require the establishment of an adequate infrastructure as a prerequisite to the equipment purchase, such as a Total Quality Management system upstream of an FMS system in order to minimize production downtime for the system (Boer, Hill and Krabbendam, 1990). In addition, some of the required additional personnel may not be used intensively throughout the whole of the implementation cycle and may be perceived as 'slack' resources (Boer, 1991) so senior managerial staff will need to be able to compare such costs against those of time slippage through unpredicted technical problems.

Although there were apparent implementation successes in the companies described in this chapter, probably because of the perceived strategic importance of the technological changes to the company's future viability and a willingness to provide sufficient resources for implementation, other publications point out that users are frequently reluctant to undergo the organizational change necessary to obtain the full benefits from integration across the various facets of design and manufacture, and departmental jealousies may hinder that change (Bullinger, 1990). In addition, the preference of engineers and technologists to divide problems into solvable segments may hinder the progress of integration across design and manufacture, particularly in view of suspicions that more holistic approaches may be impractical (Corbett, Rasmussen and Rauner, 1991).

A more recent publication reinforces the view that organizational issues are the main factors hindering the wider use of CIM, although as the cases described above reveal, hardware and software reliability are probably more significant. Forrester *et al.* (1992), having noted that the two main areas of application of computerization in manufacturing are 'engineering' (including design and pre-production) and 'production management' (including timetabling and inventory management), continue with the statement that

> CIM implementation requires mutual understanding, co-operation and co-ordination amongst all manufacturing and engineering personnel within the business. Organizational integration and the elimination of barriers have, however, proven to be more difficult to achieve in practice than technical systems integration. The companies we have studied in the course of our research illustrate severe organizational difficulties at interfaces between functional areas. Modules have been developed in isolation within different areas and only to be integrated at a later date. Attempting, then, to compel these isolated systems development teams to work together in harmony will, more often than not, pose organizational conflicts and power wrangling amongst the development personnel. In most of these companies 'interconnection' rather than integration would seem to be a more accurate description of systems development.
> (Forrester *et al.*, 1992: 10)

In the cases described in this chapter, however, it appears that companies preferred to assimilate separate sub-systems or 'solvable segments' (Corbett *et al.*, 1991), which were technically feasible in their own right, rather than follow a strategy of unnecessary integration with its added complexity and higher risk of failure. Where integration was considered necessary, the barriers to successful implementation appeared to be technological rather than organizational, although these two sets of factors may be considered as linked if insufficient Human Resources were available for problem-solving.

Forrester *et al.* (1992) continue on a more sanguine note 'that CIM can in itself be the stimulus for changes in structures, work practice and the disciplines required for integrated operation', provided that sufficient attention is devoted to organizational issues at the design and operational stages of CIM. These include a study of hierarchical frameworks where information may be limited to senior personnel, and lateral frameworks where functional areas may remain segmented and the interfacing of systems teams becomes difficult. At the operational stage, detailed appraisals are suggested for the structures, reporting systems and organizational processes, to provide the inter-disciplinary and innovative teamworking that is required by CIM.

To conclude, therefore, it is apparent that the organizational issues of CIM remain as potential areas for further research, although the technical problems should not be taken as solved, particularly when systems interface with each other and with mechanical and hydraulic components. The solution of these technical problems requires a high level of organizational skill in system specification, and a sufficient quantity and quality of Human Resource during implementation and assimilation.

ACKNOWLEDGEMENTS

The author wishes to record his thanks to Clive Handy, William Hunter, Tim Lipscomb, Phil Maxfield and Frank Reader, for their comments on a previous draft of this chapter.

REFERENCES

Ayres, R. U. (1990) *Computer Integrated Manufacturing*, Chapman & Hall, London.

Boer, H. (1991) *Organising Innovative Manufacturing Systems*, Avebury, Aldershot.

Boer, H., Hill, M. R. and Krabbendam, J. J. (1990) FMS implementation management: promise and performance, *International Journal of Operations and Production Management*, Vol. 10, no. 1, pp. 5–20.

Brödner, P. (1990) Technocentric-anthropocentric approaches: towards skill-based manufacturing, in M. Warner *et al.* (eds.) *New Technology and Manufacturing Management: Strategic Choices for Flexible Production Systems*, Wiley, Chichester, pp. 101–12.

Browne, J., Harhen, J. and Shivnan, J. (1988) *Production Management Systems: A CIM Perspective*, Addison-Wesley, Wokingham.

Bullinger, H–J. (1990) Integrated technical concepts towards the fully automated factory, in M. Warner *et al.* (eds.) op. cit., pp. 15–31.

Commission of the European Communities (1982) *ESPRIT – The Pilot Phase*, COM(82) 486 Final 1/2 CEC, Commission of the European Communities, Brussels.

Corbett, J. M., Rasmussen, L. B. and Rauner, F. (1991) *Crossing the Border: The Social and Engineering Design of Computer Integrated Manufacturing Systems*, Springer Verlag, London.

Ellis, A. and Maxfield, P. (1986) System operation, no. 4, case study, in A. Wilson and I. H. Woodward (eds.) *Computer Aided Engineering for Managers*, Open Tech Unit, Loughborough University of Technology.

Forrester, P., Hassard, J., Tang, N. and Hawksley, C. (1992) Examining computer integrated manufacturing as a determinant of strategy and organization. Paper presented at the British Academy of Management Conference, Bradford.

Harhen, J. and Browne, J. (1984) Production activity control; a key node in CIM, in H. Hubner (ed.) *Production Management Systems: Strategies and Tools for Design*, North Holland, Amsterdam.

Hill, M. R. (1985) FMS management – the scope for further research, *International Journal of Operations and Production Management*, Vol. 5, no. 3, pp. 5–20.

Hill, M. R. and Woodward, I. (1986) Managing CAE, no. 6, personnel issues, in A. Wilson and I. H. Woodward (eds.) *Computer Aided Engineering for Managers*, Open Tech Unit, Loughborough University of Technology.

Hill, M. R. and Woodward, I. (1987) *Success in Managing Advanced Manufacturing Technology*, Open Technology Group, Loughborough University of Technology.

Johnson, H. T. and Kaplan, R. S. (1987) *Relevance Lost: The Rise and Fall of Management Accounting*, Harvard Business School Press, Boston, Mass.

Kaplan, R. S. (1984) Yesterday's accounting undermines production, *Harvard Business Review*, July/August 1984, pp. 95–101.

Kaplan, R. S. (1988) Relevance gained, *Management Accounting*, Sept., pp. 38–42.

Lay, G. (1990) Strategic options for CIM integration, in M. Warner *et al.*, (eds.) op. cit., pp. 125–44.

Reader, F. J. (1986) System operation, no. 5 case study, in A. Wilson and I. H. Woodward (eds.) *Computer Aided Engineering for Managers*, Open Tech Unit, Loughborough University of Technology.

Suter, J. and Leary, V. (1989) Planning and scheduling in heavy engineering. Paper presented at Teaching Company Programme Conference, University of Newcastle-upon-Tyne, June.

Warner, M., Wobbe, W. and Brödner, P. (eds.) (1990) *New Technology and Manufacturing Management: Strategic Choices for Flexible Production Systems*, Wiley, Chichester.

8

MANUFACTURING RESOURCE PLANNING

Alan Spreadbury

Traditionally, most performance measurements within companies have been cost, efficiency, or productivity related and these are usually associated with individual departments. This situation is changing rapidly with the use of measures which focus on the effectiveness of the whole product supply chain. Manufacturing Resource Planning (MRPII) is an approach which has and will continue to play a significant part in facilitating this change and offering the framework of a formal planning and control system as a building block for controlling improvements. In mechanical terms MRPII can do little more than offer the benefits of a computerized order and material tracking system, and the journals are littered with examples of disappointed MRPII users who have gone through many of the costs of implementation but see few of the benefits (World Class International, 1991). Coupled with more enlightened HR strategies, however, which allow MRPII to be viewed for what it is, that is a powerful set of tools to be used to solve specific problems, companies can and do achieve remarkable performances. The human environment into which the MRPII system is introduced is therefore critical for success. Throughout this chapter it will be apparent that the MRPII philosophy changes many of the traditional roles in a wide variety of functions and organizational levels.

The examples cited in this chapter relate primarily to the pharmaceutical and fine chemical industry. This sector is not particularly good in 'World Class Manufacturing' terms given its characteristically high inventory levels, long lead times and only moderate customer service. Ironically, these facts place companies in this sector as likely MRPII users since the payback for those who succeed will be considerable. There are various checklists which purport to represent 'World Class Manufacturing' standard. These include the Baldridge Award and the Oliver Wight ABCD Checklist. The most important features common to such lists are measurements of inventory levels and the like.

The main example used in the chapter is Dista Products Limited, an affiliate of the American pharmaceutical company Eli Lilly. Dista is based in an industrially depressed area of south Liverpool, is a unionized site and employs around 850 people. The main business is to supply bulk animal health and pharmaceutical products to other company affiliates for formulation or direct sale, and the core processes are based on bulk fermentation, with semi-continuous or batch finishing operations.

The specific need for MRPII implementation at Dista was driven from within, not only in terms of the need for formal planning and control processes, but also in terms of the need for an integrated computer-based manufacturing system. The aim therefore was to fulfil both of these needs through an MRPII project which would also involve the purchase of new computer hardware and software. The need for a formal planning and control discipline at Dista was tested at the outset by seeking the opinions of the customers of the informal processes which were to be supplanted. The responses were clear and unambiguous: 'We get asked to make things urgently, then they sit there in the warehouse for weeks' or 'Planning always give us urgent orders on a Friday afternoon – there is a widespread belief that these have been available all week.' Production scheduling was typically carried out on an informal basis using a 'make it as soon as possible' principle until asked to change at short notice and make something else.

The need for an integrated manufacturing system which included real-time inventory data was also very pressing. All such data were held primarily on paper, or in 'red books', were always out of date and led to material planning decisons which were based largely on gut feeling or guesswork and rarely on good data. Such compelling need to modernize and redesign information flows and structures helped to generate a broadly based and enthusiastic ground swell of opinion throughout the company and no single department was seen as driving the project.

This chapter aims to describe the nature of MRPII, isolate the key features of a real implementation and summarize the benefits and impact on the organization. The sections dealing with these issues are as follows:

- What is MRPII?
- Three key stages: planning, implementation and operation
- Outcomes and impacts

WHAT IS MRPII?

MRPII has its roots in the mid-1960s with the advent of Materials Requirement Planning (MRP) which evolved from a realization that the best way to optimize and control raw materials is to formulate a needs plan based on the forward production plan rather than relying solely on historical usage analysis, statistical calculations or simple order-point 'line on the tank' methods. It quickly evolved, however, into much more than an ordering tool. Changing priorities and order due dates within the Master Production Schedule (MPS) can be reflected not only in terms of the forward material plan, but also in terms of a Capacity Requirements Plan (CRP).

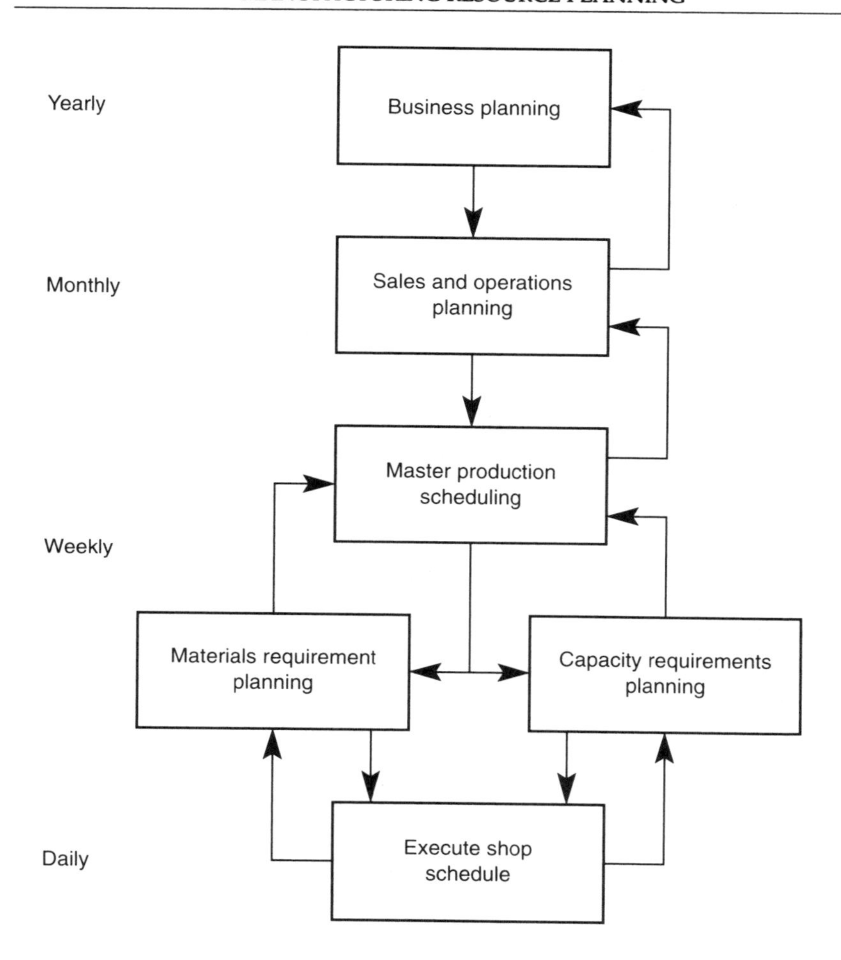

Figure 8.1 MRPII: integration of planning functions within the closed loop

Manufacturing Resource Planning (MRPII) has been defined (Wallace, 1985: 261) as 'a method for the effective planning of all resources of a manufacturing company'. Its mechanics comprise a variety of functions each linked together. These are business planning, Sales and Operations Planning (S&OP), Master Production Scheduling, Materials Requirement Planning (MRP), Capacity Requirements Planning and associated shopfloor execution systems. Figure 8.1 illustrates how these functions are integrated. In order to close the loops and provide true business integration, financial management is also included within MRPII, and this helps the business and S&OP functions to operate efficiently, through incorporation of 'what if' and financial forecasting based on 'one set of numbers'.

During interactive education sessions with shopfloor personnel who have little prior knowledge of planning and control systems, the most frequent comment received is 'Why haven't we done this before? It's just common sense.' A major fallacy which still has widespread appeal is that MRPII is simply a computerized planning and scheduling system which is often portrayed in manufacturing control circles as a purely technological rather than organization solution (Holder, 1991). As a consequence of this, implementation teams are sometimes constituted with systems rather than manufacturing expertise in mind, and the inevitable logistical problems which attend a major new software (and hardware) installation tend to reinforce the fallacy.

Other commentators have suggested that MRPII is unsuited to process industries because of the increased variability of output. What these viewpoints tend to overlook, however, is that although MRPII seeks to formalize the planning and control processes by the use of a computer-based information system, it does *not* hand over decision-making to the computer. On the contrary, key decisions concerning inventory levels, order promises, capacity and material requirements can be made with greater confidence when supported by accurate performance data and simulation (what-if) capability. This eliminates much of the 'gut feeling' element from the planning processes. It is easy to forget that MRPII (to use a popular cliché) is a 'people-driven' system and should be approached both before and after implementation with that in mind.

In every manufacturing operation, each of the basic planning functions is carried out formally or informally by someone. The question is – how well are they being done and how do you cope when a key person is absent? In other words to what degree is the company operating an *'informal system'*? The first questions which any company considering implementing MRPII should consider are:

- Are the current customer and supplier links strong?
- Are there stock-outs of product or raw materials?
- Are there missed or deferred customer orders?
- Are overtime payments excessive?
- Are the inventories excessive?
- Are the lead times long in comparison with competitors?
- Are 'hot lists', expediting and 'fire-fighting' the norm?

If the answer to many of these questions is yes, then MRPII could be a suitable tool. Good strategic planning and execution which is able to link business needs of this sort with Human Resource initiatives involving employee involvement and/or organizational change is the difference between companies which survive, thrive and ultimately excel and those which disappear.

This section describes the following 'core functions' which are integrated together to comprise MRPII:

1. business planning, Sales and Operations Planning;
2. Master Production Scheduling;
3. Materials Requirement Planning;

4. Capacity Requirements Planning;
5. shopfloor control;
6. financial integration.

1. Business, Sales and Operations Planning (S&OP)

Long-term business and strategic planning of an organization which may involve corporate dependence and sometimes remote sales and marketing input can often be a test of endurance. Each year, having spent weeks or even months requesting data and debating decisions on future capital investment, market opportunities, capacity requirements, labour needs and so forth, company directors heave a huge sigh of relief and depart on their belated summer holidays, secure in the belief that they have set the ship on course for the next twelve months. The newly 'reborn' business plan is ready to be handed down like tablets of stone to the key operating departments.

The production manager gloomily eyes the new efficiency target (never before achieved), handicapped by a freeze on headcount and a sharply-reduced plan for capital expenditure and can immediately picture next year's *post mortem*. The financial manager is still smarting after being carpeted for apparently losing an immense sum of money in adverse variances or inventory write-offs over the previous year and wonders why no one ever listens to financial advice except in the run up to business planning. The planning manager is just glad it is all over; he can get back to the real job of trying to balance what the factory produces with the demands of those awkward customers. He ponders, however, why the production manager won't speak to him.

This is an all too common problem. The plan is often prefaced by nasty surprises, and dropped like a hot potato into the laps of line managers. MRPII philosophy is both specific and prescriptive about how the business and strategic planning functions need to be organized in order that opportunities are not missed, surprises are few and decisions are made without the need to toss any coins.

Sales and Operations Planning (S&OP) is for many companies a completely new planning process, which replaces an informal planning function usually owned exclusively by experienced people with the planning department and operated (some would say) by people as remote from production as the Earth is from Mars!

The complete business plan in a manufacturing company will include R&D and other expenses not directly related to production, but at the core of the business plan of any MRPII operation is the S&OP. As the S&OP is updated each month, it is translated into money, and the business plan progress is thus monitored regularly. Each month the S&OP considers for each related family of products the sales plan, production plan and anticipated inventory holding. The planning horizon usually extends to two years, and time 'fences', set in accordance with manufacturing and other lead times, are used to ensure maximum short-term stability of the plans. A

End of month close-off	Sales and marketing pre-S&OP meeting	Manufacturing pre-S&OP meeting	S&OP meeting	Prepare detailed MPS plans
Update financial, sales, inventory and manufacturing plans Measure and report performance against set plans	Review sales and forecast performance for previous period Highlight changes in demand (volume, mix or timing) and consider impact Analyse assumptions and vulnerabilities and make requests for changes to manufacturing or inventory plans	Review last month's manufacturing performance Consider any proposed manufacturing or inventory changes Evaluate capacity (detailed or rough-cut) and material requirements to support the new plan Cost all realistic options and prepare recommendations	Review last month's sales, manufacturing, inventory and financial performance Consider any new recommendations to support the new plans and decide actions based on realistic expectations Authorize changes to medium- to long-term future plans and fix the manufacturing plans within the 'firm' period	Reconcile Master Production Schedules with new manufacturing plan Communicate detailed plans to operating departments

Figure 8.2 Sales and Operations Planning cycle in an MRPII environment (time-scale one to two weeks)

typical S&OP cycle is shown in Figure 8.2. To facilitate easy access to manufacturing performance data, there needs to be a single set of numbers and not a financial manager's version of what has been produced versus planning's versus production's. What Figure 8.2 shows is that the sales and operatons meeting is, or rather should be, only one point in a more continuous planning cycle.

The Dista planning process formerly consisted of a fortnightly meeting usually attended only by managers and department heads from the planning and production area. Much of the meeting was spent debating how much product had actually been produced and how much was tested, approved and ready to be sold. Forward production plans tended to be limited to the next fortnight, and were often changed within this period at significant extra set-up or clean-down cost. This process was highly inefficient, often caused unnecessary changes at short notice, and prompted much frustration. Strategic investment decisions were also inhibited because of a lack of confidence in future projections, and the absence of an integrated data collection and reporting system resulted in much labour-intensive preparation for each planning meeting.

The current S&OP process at Dista now involves only executives or senior managers from key operating departments. Monthly performance in respect of production, sales and inventory is reviewed at family level, and forward plans stretching out eighteen months are approved. Progress on strategic outsourcing or new projects is checked, as is adherence to the business plan.

2. Master Production Scheduling (MPS)

In the closed-loop MRPII system, monthly production plans for family groupings approved at the S&OP meeting are translated into discrete weekly requirements and further divided into individual end item needs. The resultant master schedules are pushed forward as the firm planning horizon is extended week by week and 'suggested orders' are made firm. The master schedule is the linchpin in the MRPII system, and the effectiveness of the master scheduler is critical in this environment. In addition to his role in order promising and S&OP support, he will spend a very significant part of each day communicating formally or informally with manufacturing. This half planning, half production role is in practice a very challenging one, and demanding of excellent communciation skills as well as technical knowledge and confidence. Figure 8.3 illustrates the skill which the master scheduler (a new role for most companies implementing MRPII) must demonstrate in balancing supply and demand factors.

This type of skill and experience is unlikely to be found in many of the existing encumbents of the planning department. A production background is probably the best preparation for master scheduling, and a good master scheduler will be equally at ease whether evaluating with a demand planner a change in the sales forecast or discussing potential schedule changes with shopfloor process operators. Once master schedules are created and

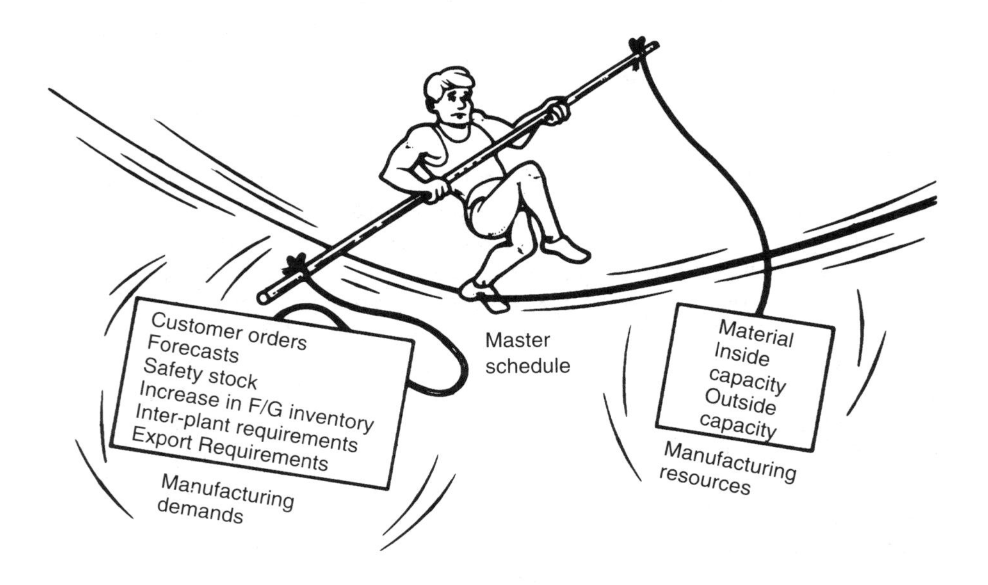

Figure 8.3 The master schedule as the balance between manufacturing demands and manufacturing resources

capacity and material needs are reconciled, the master scheduler creates detailed daily schedules which are typically communicated electronically by means of a dispatch list which can be updated by shopfloor people.

Dista's experience with master scheduling was that progress during implementation was slowed considerably by the limitations of the MRPII software, which required extensive functional enhancement. Also, after changing to the new planning system, there was significant schedule instability caused by some out-of-control processes. This caused major frustration for the master schedulers who become known for a time as 'master reschedulers'.

As the new processes gradually settle down and become accepted, the 'planners' of the old organization (whose main task was *expediting* rather than *planning*) become allocation or customer service personnel, and the order-promising people begin to find less call on their creativity in finding excuses when the customer calls. This has certainly been the experience at Dista and, gradually, old card-index systems can be dispensed with as confidence in the single set of electronically-held numbers grows. Complaints about the new system may at first be frequent, but calls to return to the old way will be few.

3. Materials Requirement Planning (MRP)

Materials Requirement Planning (MRP) in effect begins by asking the 'universal manufacturing questions' listed in Figure 8.4.

• What are we going to make?	Master Production Schedule
• What does it take to make it?	Bill of materials
• What do we have in stock?	Inventory records
• What do we need to get?	Materials Requirement Plan

Figure 8.4 The universal manufacturing equation

The simple arithmetic linking these questions through the functions on the right side of Figure 8.4 is standard logic repeated at least weekly, and often daily, within any MRPII system. For manufacturing operations which have typically operated either informal raw material procurement practices or formal or informal order-point systems, MRP offers several benefits. The first and most obvious benefit is financial. Optimization of stock levels linked by MRP to the MPS will ensure that key raw materials are purchased in correct quantities to arrive in a timely fashion, usually resulting in inventory savings. Visibility of forward product requirements of up to two years can be easily translated into supplier schedules giving selected vendors similar forward visibility with attendant scope for financially favourable contracts.

The second type of benefit which can be expected is a significant realignment of job descriptions within the purchasing function. Buyers who currently spend time expediting items of both high and low value (often the latter being the most preponderant) will, with greater confidence in medium- to long-term requirements, spend more time negotiating better contracts with key strategic or 'high spend' vendors. Supplier scheduling will probably be introduced as a separate function and will link directly with suppliers to maintain and manage valid delivery schedules. Both these activities have become pivotal in building strong new bonds with outside suppliers through supplier certification processes – a far cry from the days when suppliers were treated with disdain and played off against one another. Leading-edge companies now invest significant systems or problem-solving resources in improving the manufacturing capabilities of their suppliers.

One might imagine that the impressive electronic corporate communication tools now widely in use and the 'common corporate culture' effect would facilitate the building of strong links between affiliates who supply each other, but in practice the opposite is often the case. All too often, dual standards are applied to inter-affiliate business, and objective performance indicators show that 'non-captive' but closely linked suppliers perform much better. Where inter-affiliate transactions are used, they should be on a strictly monitored and impartial basis.

An essential point to emphasize with respect to MRP systems is that in order to gain any of the considerable benefits, the amount of error correction and dealing with exceptions must be minimized. Specifically, significant instability in the master schedule or inaccuracies in bills of material or inventory records will result in erroneous material requirements and inordinate amounts of time wasted correcting errors or dealing with changes. It is Dista's experience that until this was addressed effectively, most of the benefits outlined above were negated. In the face of poor quality data, the buyers and supplier schedulers will either revert to type and use the old informal 'grapevine' or 'finger in the air' methods of forward planning or, worse, there will be excessive inventories built to guard against stock-outs. The intrinsic importance of factors such as bill of materials and inventory accuracy within the MRP environment can also yield spin-off benefits in terms of compliance with regulatory bodies' mandatory codes of Good Manufacturing Practice (GMP) for pharmaceutical or related industries where the need for lot control and traceability increasingly demands the highest standards of material management and control.

4. Capacity Requirements Planning (CRP)

The applicability (and hence the priority during implementation) of this technique to an operation will depend on many factors, but operations which have work centres producing many different types of product will be the main beneficiaries. As with any forward planning technique, CRP is limited by the accuracy of forecasted demand, and the stability of production schedules, but, as is illustrated in Figure 8.5, once a good CRP is available, provision of capacity through the use of extra shifts, movement of labour, overtime and revision of the master schedule can be considered in a timely and measured way. This contrasts with the use of an informal system in which any increase in required capacity inevitably necessitates short-notice overtime payment or similar unwelcome 'surprise' for management.

Without a CRP you are in danger of always undertaking short-term costly fixes. The line drawn vertically in Figure 8.5 after week 8 is the so-called

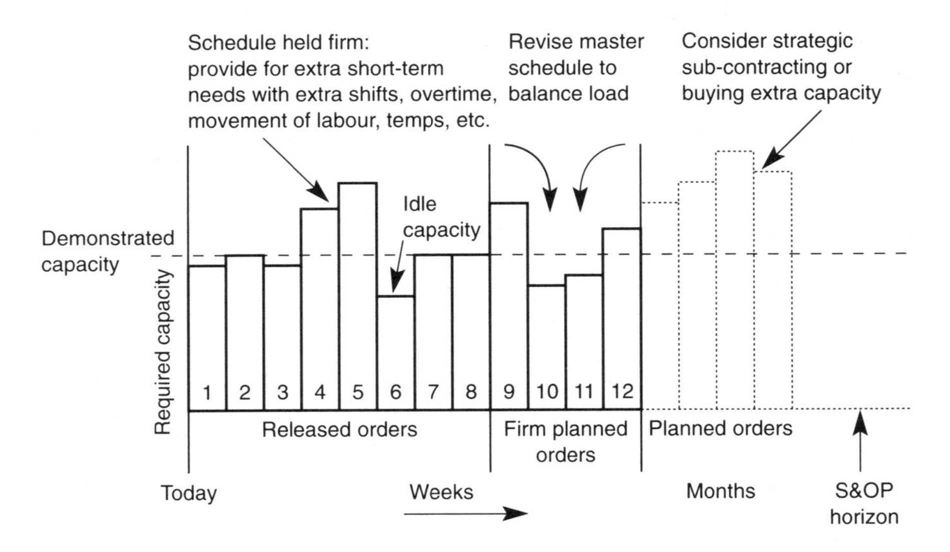

Figure 8.5 A simplified Capacity Requirements Plan

'firm fence' – that is, a boundary within which one tries very hard not to make any changes to the schedule. CRP is driven directly by the Master Production Schedule. Just as the bill of materials is combined with the master schedule to give the detailed forward Materials Requirement Plan, the product routeing, which details work centre loadings, is used in conjunction with the master schedule to calculate a detailed forward capacity plan. Both released and planned orders in the schedule are used to generate a capacity plan for each work centre, and a simplified form is shown in Figure 8.5. Because CRP calculations use logic in which a work centre can be infinitely loaded, the required capacity can sometimes exceed the demonstrated capacity. For this reason it has been attacked by critics (Holder, 1991) who are uncomfortable with the idea of labour being planned in a similar way to material. This is a facile argument, since CRP will only find optimum use in those organizations which are flexible in the way they manage their Human Resources and this ideally should not be a constraint on the process of producing the right products on time. It simply emphasizes the importance of and need for complementary and synergistic planning and Human Resource (HR) policies.

Unfortunately, for many companies, changes in the flexibility of labour lag behind advances in planning and scheduling and so constraints still exist. Any shortage (or excess) in capacity can be met with either a change to the schedule or an adjustment in the capacity itself. The absence of effective detailed capacity planning where it is appropriate invariably leads to capacity mismatches not being predicted, and expensive short-notice overtime, idle labour or unsatisfied customers will result. The development of CRP as a tool within the MRPII closed-loop system forces a company to develop policies on how to deal with capacity issues, particularly regarding Human Resources, and supports the thesis that *the problem of managing capacity is far less one of technical control than of people management*. Although every MRPII computer software will produce detailed CRP reports, the power of this technique is not that it makes decisions for management, but that it puts the alternatives clearly into focus so that better human decisions can be made in advance.

5. Shopfloor control

Having put in place a formal infrastructure of synchronized planning activities, beginning with a view of whole product families over a three- to five-year horizon, and ending with daily shop schedules for individual items, efficient and timely feedback of manufacturing execution progress is essential to provide information for the closed-loop process to continue. In practice, the nature of the shopfloor linkage will vary widely from plant to plant, and will range from automated electronic data capture within assembly line, Just-in-Time (JIT) or flow environments to *kanban* systems or manual job card entries.

Since this aspect of the control system is likely to involve the highest proportion of the workforce, careful thought needs to be given to: (1) the HR implications of any new technology, and (2) how any proposed MRPII implementation fits in with current initiatives. At Dista, many of the production areas formerly operated from informal word-of-mouth schedules, and due dates for jobs were almost unheard of. No manufacturing control system existed, and a high proportion of shopfloor personnel had no experience of terminals or keyboards. Against the background of this huge training task, an early decision was taken that each process operator would be responsible for entering real-time progress and material usage data rather than using clerks to carry out this function after the fact. The shopfloor responded positively to the introduction of these new methods, showing the value of the emphasis placed during training on explaining 'why' as well as 'how to'. It is hard to estimate the impact of this decision, but it is certainly true that many companies opt to limit the number of people entering data in order to minimize errors.

A significant challenge at Dista was to integrate blue-collar operations with white-collar testing and quality control (QC) functions into a single formal chain of operations. A perception of testing and QC people as 'policemen', together with differences in shift versus day operation between the two groups, was a serious barrier to progress. In spite of this, however, the decision to make them interdependent on product routeings within the MRPII system resulted in a pressure which continues slowly but perceptively to draw them together. With the advent of valid due dates for each operation, development of these links is easy to measure and over a period of time expediting and 'hot lists' gradually disappear.

6. Financial integration

The importance of this to a successful MRPII system is sometimes overlooked. Valid long-term sales and production plans, efficient daily shopfloor scheduling and accurate and timely material issues and receipts need to be translated from kilos or other operational units into sterling so that cashflow and financial forecasting is effectively underpinned. For many companies, Dista included, the end of the month meant a paperchase followed by dozens of phone calls to check clarity or accuracy of data, then laborious keying in of raw data into a PC or stand-alone computer system. Reconciliation of costing variances was relatively easy because most of the checking had been done at the raw data input level.

It is very important to choose MRPII software which takes into account the need to support the business financially and it is critical to involve financial people intimately with the project from the outset. Many companies find this difficult and tend to regard finance as a monthly irritation rather than an essential support for the manufacturing supply chain.

Three elements of financial activity which are very important to integrate with the manfuacturing control system are production costing, standard costing and accounts payable. Computer interfacing of costing through realtime order and material usage tracking potentially allows the automation of a once extremely labour-intensive task amd frees financial staff for more value-adding activities such as financial forecasting and budgeting, financial budgeting and guidance on priorities. Similar logic justifies a need to simplify and integrate the three-way matching process between purchase orders, order receipt and payment of accounts. The experience at Dista was that realization of these benefits was a slow process. Although the financial data, now integrated, are more easily accessible, they are also much more susceptible to errors. Because the manufacturing/costing data link is now present within the computer, errors in issuing or receiving materials cannot be easily 'screened out' as was the case during the manual data entry. A similar situation exists within accounts payable in respect of errors made during ordering or receipt of purchased goods. The full benefits of financial integration are therefore contingent on working with the now greatly enlarged population who enter data to find new and better ways of ensuring its integrity.

PLANNING, IMPLEMENTATION AND OPERATION

There are a few key phases in the lifetime of an MRPII implementation, and each has to be considered discretely in terms of what resources are required, and what intermediate targets should be in focus at each stage. These are summarized in Figure 8.6. *Each is associated with a set of organizational and Human Resource problems and considerations.*

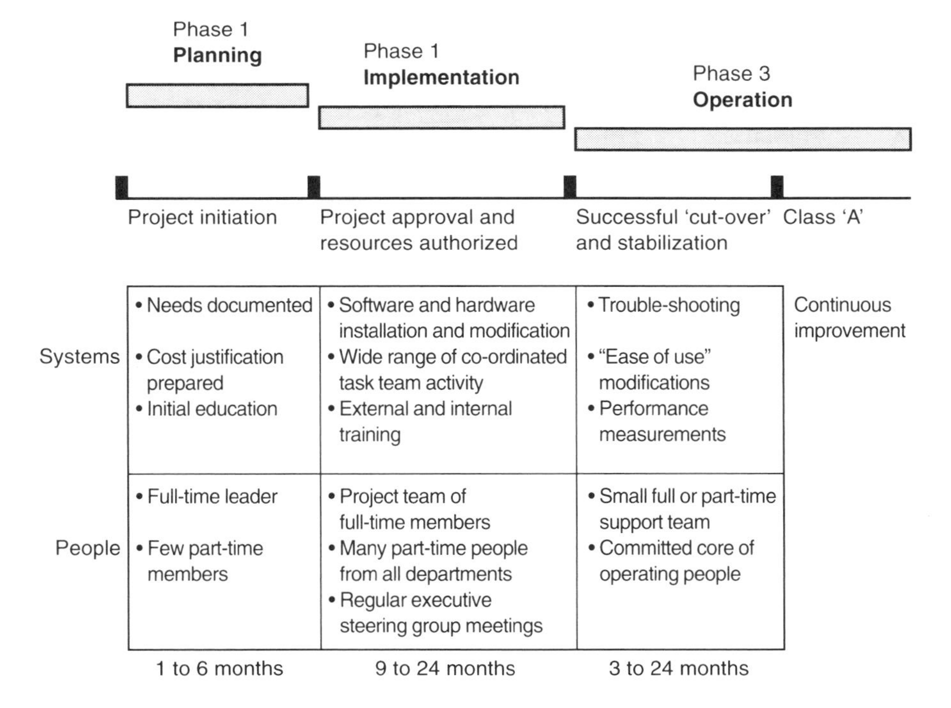

Figure 8.6 MRPII project lifetime

For example, in the first six months the core activity entails cost justification and initial education of those people who are driving the MPRII project. The associated people problems are getting high-quality people sufficiently committed and ensuring top management support. In the second phase (nine to twenty-four months) task teams are established and the organizational and people problems relate to co-ordination and control. At this time the MRPII project manager has to cajole and yet also orchestrate a large number of people and diverse task teams – many of which are likely to be interdependent. In the third phase when the project has 'gone live', the typical technical problems hinge on system trouble-shooting relating to data error correction and the associated people problem is to ensure adequate motivation to sustain the discipline which is needed to maintain a formal system (see also Burns and Turnipseed, 1991).

It is unrealistic to expect an MRPII implementation project to continue as the 'number one' focus of an organization for a period of much more than about two years, and any one of a number of factors in practice is likely to influence the time-scale and perhaps ultimately the overall success or failure. These will include some or all of the following:

- changes in key project personnel;
- changes within senior management structure;
- external influences, e.g. regulatory;
- changes to business goals, e.g. new product or project launches;
- changes in HR utilization, e.g. organizational restructuring.

What is critical to success in the face of these changes, however, *is the resolute and sustained support of influential individuals at all levels*. The project needs its ambassadors at each level in the organization, and erosion of this influence at board level in particular is usually the signal to abandon ship. Conversely, the appointment of an enthusiastic chief executive officer is often the fillip which breaks the stagnation and restarts an ailing project.

There is sometimes an unwillingness on the part of top management to listen to 'gurus', particularly where organizational change, relocation of staff or redefinition of roles is being counselled. The types of change accompanying an MRPII implementation often represent a fundamental challenge to traditional departmental 'silo' mentality and impact not just the technical content of some jobs, but also reporting and other cross-functional relationships. For example, the introduction of the master scheduling role was problematical during the Dista implementation. The master schedulers were intended to have a formal role within manufacturing and material supply as well as planning. Who should they be? Who should they report to? Where would they sit? What would they do? What salary scale would they be on? All these issues took an age to sort out and were a warning signal that any fundamental change to the organizational structure, which had, after all, evolved over decades, would require a certain amount of faith and conviction to initiate. Other similar resource decisions prompted during the implementation are whether to provide supplier schedulers, material planners,

S&OP analysts and capacity planners. This type of decision will rarely be made at the initiation of the project, because there is unlikely to be any in-house experience of such matters and any local headcount increases will always be unpopular and may consume much management time to resolve.

Planning stage

The inception of the project will usually involve a small number of influential managers or executives becoming familiar with MRPII philosophy at an overview session provided by an external consultant. One full-time person or a small number of part-timers will then be expected to prepare a cost-benefit analysis and to examine in broad terms the scope for improvement. The project leader should be someone with several years experience with the company, preferably from one of the key manufacturing departments, and should be someone who commands respect and credibility.

This cost-benefit analysis will often include the purchase of new computer software and/or hardware, as well as provision for education, consultancy and other associated costs. At Dista, for example, a total project cost of just over £1 million was budgeted on a projected payback of five years, predominantly through savings in inventory holdings. This type of visible and tangible justification, however inaccurate, is important to gain initial support for the project. As well as implementation costs and cash-flow analysis, the HR requirements for the project will be significant. Overall, headcount reduction is unlikely to be a major factor in the justification of the project, but during implementation the demands are very significant and need to be budgeted in advance. In addition to the full-time project leader, other full-time and part-time people will be required. For example, at Dista, a site with approximately 850 employees, a total of five full-time project members were used for a period of around two years, extending past cut-over into the operation phase. In addition, approximately five other ambassadors led task teams on a part-time basis, these involving a peak of around fifty people during six months prior to going 'live'. Despite this very significant effort, at least one further full-time systems analyst was contributed through inter-affiliate system support. At this stage the potential changes to the organization need to be 'sanity checked' with a wide cross-section of functional areas, and at this time some allies will emerge from the list of the biggest potential beneficiaries.

Implementation stage

Once funding is approved, more people need to become active on a part-time basis, and a full-time project team should be assembled. The time-scale for the implementation stage will obviously vary according to the level of resources assigned. Selective external education is usually used to help create the necessary ambassadors within all key operating functions and to populate the spin-off task forces which will be needed. These will cover some or all of the following topics:

- system implementation and software enhancement;
- bills of material/static item data;
- shopfloor routeings and lead times;
- Sales and Operations Planning;
- Master Production Scheduling;
- inventory records/data accuracy;
- financial integration;
- purchasing/supplier scheduling.

Each taskforce activity requires organization and prioritization, and interdependent tasks need effective co-ordination. Two methods of doing this utilized with some success at Dista were the use of off-site 'total project' meetings with all full- and part-time people present and weekly manager meetings during the few months prior to going 'live'.

The initiation of an S&OP process is often the first foundation stone which is put in place during the early stages of the project. This requires little or no computer support, and can set strong direction for the rest of the task teams. It can also provide a regular high-level forum to view how well the business is being controlled by managing sales, production and inventory. At Dista, the formal involvement of a senior sales person at the S&OP meeting also became an essential ingredient for success. This had been absent in the pre-MRPII days because the sales force operated from another site.

Depending on the product range involved, it may be valuable to pilot some or all of the software and MRPII functions within a 'slice' of the business prior to full-scale utilization. The objectives here would be to demonstrate some early payback as well as generating enthusiasm for some of the new tools. At Dista one of the main benefits of this was to distinguish between essential and nice-to-have software enhancements. The 'slice' approach was less effective at Dista in demonstrating some of the other key functions such as master scheduling and Materials Requirement Planning because at that stage in the project the necessary organizational changes had not been made, and parallel operation of MRPII methods for one product line with 'old-style' methods for the rest created unacceptable pressures for some individuals.

One of the major activities in terms of effort during the implementation phase is likely to be training and education. Up to this point, most of the full- and part-time implementers as well as many task team members will have attended external MRPII classes specific to their area of expertise. Before full implementation, everyone involved needs both system training and MRPII education, and it is the job of the project team to set this up by designing internal system training and MRPII education courses tailored to the specific needs of the workforce. These should be as interactive as possible, and for maximum effect, need to be delivered as close as possible to full operation of the system. This proved to be a major logistical challenge at Dista. Courses were organized such that full-time project team members delivered the educational elements and line management delivered the hands-on training.

Prior to going 'live', more than 200 people received approximately a week of training comprising both MRPII education and hands-on computer training. Apart from the organizational challenge this presented, the major shortcoming of the plan was that training was carried out using test and simulated data prior to going 'live' in production, and inevitably weeks or even months elapsed between training and actual usage. The result was that despite the very large training effort, it still took about six months to stabilize the business because of elementary keying errors necessitating error correction and trouble-shooting.

At every stage during implementation there will be a steady stream of software improvement tasks. These may have been prioritized into critical and non-critical, and in the weeks immediately prior to cut-over the analysts and programmers will feel the greatest strain. None of the commercially available software has captured a major market share, and this is because the software developers have opted for the revenue from support and product enhancement contracts rather than investing heavily in product development to produce a solution which can be implemented off the shelf with little or no modification. In reality most of the available MRPII software requires significant customization in order to (a) fit with the specific needs of an individual company and/or (b) provide basic MRPII functionality which may be (and often is) absent. There are dozens of MRPII software packages, most of them designed with particular industries in mind. The central point to note is that it is impossible to buy off the shelf a system that can be used uncustomized. Major, and costly, customization is usually required. Systems sales people rarely make this point sufficiently clear.

In practice, the software implementation issues are usually irritating and distracting; they can slow the project down dramatically, but they are not usually among the most critical success factors. During implementation, the software focus is on providing basic functionality, whereas after going 'live', the emphasis begins to shift towards factors such as user friendliness, ease of access, ability to replace tried and trusted manual paper systems and ability to provide performance data in a clear and conventional format. The software producers of the future will ignore these post-implementation requirements at their peril.

Operation stage

The first major milestone has been passed when successful implementation, defined as implementation of core MRPII processes and stable computer systems, is complete. These must not be critically dependent on old informal parallel systems, and must not consume inordinate resources just to keep running. When this point is reached, a change of emphasis needs to be consciously managed by the steering group. When implementation is complete, successful operation needs to be diverted towards a set of now universally accepted standards for excellence, sometimes called MRPII Class 'A'. For a variety of reasons, many companies lose focus at this staging point. For

1. Management commitment

Planning and control processes

2. Strategic planning
3. Business planning
4. Sales and Operations Planning
5. Single set of numbers
6. 'What if' simulations
7. Forecasts that are measured
8. Sales plans
9. Integrated customer order entry and promising
10. Master Production Scheduling
11. Supplier planning and control
12. Material planning and control
13. Capacity planning and control
14. New product development
15. Engineering integrated
16. Distribution resource planning

Data management

17. Integrated bills and routeings
18. Bill accuracy 98–100%
    Routeing 95–100%
    Inventory records 95–100%
19. Product change control

Continuous improvement

20. Employee education
21. Employee involvement
22. One less at a time (continuous improvement)
23. Total Quality improvement process
24. Product development strategy
25. Partner relationship with customers

Performance measurements

26. Production planning performance ± 2% of plan
27. Master Production Schedule performance 95–100%
28. Manufacturing schedule performance 95–100%
29. Engineering schedule performance 95–100%
30. Supplier delivery performance 95–100%

Company performance measurements

31. Customer service delivery to promise 95–100%
32. Quality performance measured
33. Cost performance measured
34. Velocity performance measured
35. Management uses measurements for improvements

Figure 8.7 Oliver Wight ABCD Checklist. (Source: Oliver Wight, 1989)

some companies, two, three or more years may have been consumed during implementation. Another initiative or external stimulus may be a higher priority or an influential supporter may have moved on. Other common attention grabbers within the pharmaceutical industry may be, for example, the urgent need to respond to some regulatory stimulus involving an environmental or quality improvement project. In many other cases, the lack of early payback will have exhausted patience with MRPII as the 'way ahead'. The more successful companies, however, will have learned how to integrate MRPII in such a way that is not seen as a competitor to other essential change programmes, most of which are covered in other chapters of this book.

The key areas covered by the famous Oliver Wight ABCD Checklist are shown in Figure 8.7. The items on this list, which include issues such as employee involvement and employee education, usually come as a surprise for those who previously viewed MRPII as a glorified computerized planning tool which could be bought 'off the shelf'. In fact closer examination of the checklist reveals how wide is the scope and how exacting the standards. One strategy for achieving compliance across a broad front following successful implementation and one which has been used to good effect at several companies (e.g. ICI, Kodak, Dista) is to assign a manager or director as 'owner' for each checklist question and to monitor improvement progress regularly at the executive steering group forum. Many companies also localize the checklists and adopt in-house terminology to articulate the requirements more clearly. The Oliver Wight ABCD Checklist shown in Figure 8.7 contains four main sections (planning and control processes, data management, continuous improvement and performance measurement), and each of the thirty-five overview questions is underpinned with detailed audit questions. It is also noteworthy that the checklist includes questions dealing with Total Quality and communications.

The overall aim of most, if not all, MRPII implementations is to achieve a payback through attainment of the excellent standards within the ABCD Checklist. These include >95% on time delivery, >95% manufacturing performance, improved inventory turnover, less waste, etc. The magnitude of this task in relation to the resources assigned to carry it out sometimes means that projects take inordinate lengths of time and lose momentum. Despite this, however, there are many European companies which have reached the standard within the last few years. These include various affiliates of Cyanamid, Kodak, ICI, Coca Cola, and Engelhard, as well as Parke Davis, Reckitt and Colman, Formica and Courtaulds Films, to name a few.

The celebrations which follow the implementation of the new system can soon give way to a sense of anticlimax or disappointment if visible benefits are not immediately apparent. In fact, preoccupation with things like transaction errors and data accuracy consume large amounts of time and 'fire-fighting' and reacting can become the norm. This critical stage must be managed purposefully after implementation is complete without losing sight of the next goal which is performance improvement.

An impartial assessment is needed to establish priority areas for improvement and, for the most part, these activities will now not be under the umbrella of the MRPII project. Rather, they must be driven and managed within the mainstream organizational structures. The MRPII project members, if not already assimilated into other functions, will become facilitators aiming to provide stimulus and technical knowledge for performance improvement. For the most part the planning and control processes will be in place (although perhaps not yet well documented and proceduralized), and data management in terms of accuracy of static and dynamic data should be good.

Continuous improvement in this context focuses on things such as building new and better customer and supplier links. There will be emphasis, perhaps for the first time, on treating customers and suppliers like partners and sharing with them the vision for the future. This area also overlaps with the many employee involvement and/or TQM initiatives which seem to be present in virtually every company, and the main challenge is to weave these together into some form of coherent plan instead of a set of disparate and individually championed initiatives. Once implemented, it may be expedient to focus on one family or product line and start pushing for significant improvement there. If this is not done, the magnitude of the task when viewed across all fronts may be very daunting indeed.

For most companies the advent of MRPII will bring new performance indicators, and there is real danger of submerging in a sea of paper and numbers. The strategy for dealing with this requires careful thought, and solutions can be technological, for example by providing on-line user-friendly access to data, or human, by providing a full-time data management function.

The mushrooming of new indicators with which to measure business performance, post-MRPII cut-over, offers many opportunities previously unavailable to measure individuals' performance. Some examples are listed in Figure 8.8.

There was little difficulty in gaining acceptance for the implementation of these at Dista. The management of so many performance indicators, both internal and external, provides new opportunities to pinpoint root causes of problems and often supports a cross-functional rather than departmental approach to problem-solving. Because manufacturing processes are organized, planned and tracked as connected stages (or work centres) on a product routeing, interdependency between internal suppliers and customers is more obvious. For example, if a work centre breaks down or causes a delay, there is an obligation to formally advise both the master scheduler and the downstream work centre. Formerly, this would have been placed in the

Demand planner	Forecast accuracy
Master scheduler	Customer service No. of schedule changes/delays
Shopfloor worker	Work centre on-time conformance Lead time conformance
Technical engineering	Routeing/data accuracy
Buyer/supplier scheduler	Supplier performance

Figure 8.8 Post-MRPII measurement opportunities

hands of supervision to expedite and communicate the delay. These changes are not momentous, but support gradual movement away from vertical functional management towards horizontal customer-driven product flow management.

OUTCOMES AND IMPACTS

At the very start of this chapter it was pointed out that MRPII, unlike most ordinary performance management systems offers a systematic and logical set of tools which acts upon and can lift the performance of the whole supply chain rather than merely effect incremental improvements in individual departments. It was also emphasized, however, that if treated as a mere mechanical fix, MRPII systems can be expected to fail to meet expectations. Only when MRPII methodology is handled within the context of an appropriate HR strategy does it become a truly powerful tool for change. It can be postulated that the high failure rate of MRPII implementation (Rayner, 1988) owes a great deal to the neglect of this fact.

In this final section we re-emphasize the importance of this point. Drawing largely upon the case of Dista, the case which has been used in this chapter, a number of the critical success factors can be summarily listed:

- top management commitment throughout;
- ability to integrate other change programmes synergistically;
- make the computer work for the people and not vice versa;
- make timely organizational changes to match changing needs;
- use credible and enthusiastic implementers;
- promote greater employee involvement;
- commit best resources to both training and education.

Top management must have a strategy which links the various change programmes together. Normally, numerous changes are introduced on a serial basis. Managers and employees have difficulty in seeing any connection between the different initiatives. It is also vitally important to commit sufficient resources to education and training if the benefits of MRPII are to be gained. Extensive training was undertaken at Dista and here as well as in other cases cited, the lesson has been that even more was needed.

When discussing critical success factors it is also important to note what is perhaps the most fundamental point of all. *MRPII, no matter how well it is implemented, will not compensate for lack of attention to underlying problems such as lack of investment in plant, equipment and quality systems.*

In terms of outcomes, the importance of the organizational and Human Resource issues associated with MRPII implementation can again be readily seen. For example, at Dista there was a major impact on the changing roles of individuals, groups and functions at all levels. Some of the key ones are summarized in Figure 8.9.

For those corporations able to find the right formula for MRPII to work, there are significant benefits to be gained by diffusing best practice across

Shopfloor	Operators become an essential part of the manufacturing information feedback loop Enhances skills at shopfloor level; people valued for more than simply manual output
Warehouse	No longer the place to put people who are unable to operate a manufacturing process Need for inventory accuracy given a high profile and this helps to build self-esteem within stores operations
Planning	Newly created position of master scheduler; one which is critical to the successful operation of the business Planners of the old organization become allocation personnel Customer service staff do less expediting and are able to focus on getting closer to the customer
Testing/quality control	Shift in mind-set away from departmental efficiency towards process improvement encouraged Becomes an essential link in the manufacturing chain rather than a 'policeman'
Purchasing	New supplier scheduling role created releasing buyers to help promote partnership approach through supplier and part certification Fewer suppliers with longer-term contracts means increased responsibility for buyers
Financial	Become part of the planning process each month through budgeting and forecasting with a clear role in helping to ensure the business plan is met Movement away from role as harbinger of periodic after-the-fact bad news
Executive group	Involvement in the monthly S&OP process helps to ensure that timely strategic decisions are made by choice rather than by default; fewer surprises increases willingness to delegate and improves trust

Figure 8.9 Some changing roles at Dista resulting from MRPII

other businesses within the group. Following initial success in achieving Class 'A' at its pharmaceutical and paints sites, ICI has embarked on an ambitious worldwide roll-out programme which involves sharing of site expertise. Similarly, success for Kodak at Kirkby has been followed by Class 'A' at its Annesley plant.

These and all other Class 'A' companies report the following benefits, both tangible and intangible. They continue to improve as other companies are wilting under harsh competition and they continue to attract investment at a time when this is scarce.

- Improved and formal strategic planning process (S&OP) based on sound data facilitates decision-making and enables core business elements of sales, production and inventory to be planned and controlled effectively.
- Realistic promising coupled with better manufacturing control gives improved customer service.
- Allows controlled reduction of inventory at finished goods, Work in Progress and raw material level without adversely affecting customer

service. Reduced inventory and/or improved sales result in shorter inventory residence time.

- Allows better forward planning of new products, project installations and planned maintenance.
- Enhances employee involvement through on-line availability of valid priority shop schedules, and strengthens internal customer supplier chains.
- Encourages stronger and more reliable partnerships with main suppliers through supplier scheduling and qualification/certification programmes.
- Gives better raw material control through Materials Requirement Planning resulting in fewer surprises and fewer stock-outs.
- Improves responsiveness and flexibility through lead time reduction and helps to support elimination of non-value-adding activities (JIT).
- Integrated financial elements (one set of numbers) give more accurate and timely data, create fewer surprises and promote better business planning and forecasting.
- Formal manufacturing controls (both computer and human) support the principle of the 'quality system', and enhance drives toward ISO 9000 certification.

Clearly an MRPII implementation driven towards Class 'A' which embraces all the elements of improvement outlined above, and others listed in the ABCD Checklist (Figure 8.7 above), is more than just a planning and scheduling system, and the distinction between an MRPII user company and an MRPII Class 'A' company is stark. In the final analysis though, it is not the banner which is important (and 'MRPII' is a pretty uninspiring banner at that!) but the results. There may come a point – perhaps soon after going 'live' when the new methods of working are well established – when the banner becomes more of a hindrance than a help. Perhaps the MRPII tools can be used to good effect to support and underpin other necessary TQM initiatives. Such is the overlap of underlying principles and supporting behaviours that the name hardly matters as long as there is a clear understanding of what the vision for the future is and which path is leading towards it.

There is a further lesson. Although the commercial packages and their associated checklists imply a wide-ranging programme of change, the core prescriptions associated with MRPII are heavy on planning and control issues and light on the other supportive items which appear on the checklists.

It is necessary therefore to end on a highly realistic note. The path from implementation to Class 'A' is usually a long one. Most, if not indeed all, of the Class 'A' companies in the pharmaceutical and fine chemical field took more than three years to achieve it. Payback may not occur for five or six years. Even intermediate measures of results may not always be immediately forthcoming. In some case companies looked at, inventory levels actually *increased* (in fact they doubled) two years after implementation of

MRPII. MRPII is not a panacea. Quality problems can prevent the meeting of schedules no matter how sophisticated the plan. No one should be in any doubt that MRPII is a valuable tool but its effective use, as the Dista case and the analysis in this chapter have demonstrated, depends crucially upon handling the organizational and Human Resource aspects in a professional and effective manner.

REFERENCES

Burns, O. M. and Turnipseed, D. (1991) Critical success factors in manufacturing resource planning implementation. *International Journal of Operations and Production Management*, Vol. 4, pp. 5–19.

Holder, R. (1991a) MRPII: misused or a flawed philosophy? *Engineering Computers*, May, pp. 43–8.

Rayner, K. (1988) Implementing MRPII: is it a dream or a nightmare? *Management Accounting*, Vol. 66, no. 5, pp. 25–27.

Wallace, T. F. (1985) *MRPII: Making it Happen*, Oliver Wight Co., Essex Junction, Vermont.

Wight, O. (1989) *The ABCD Checklist*, Oliver Wight Co., Essex Junction, Vermont.

World Class International (1991) *A World Class Report on the Results of a Benchmark Survey of Pharmaceutical and Fine Chemical Industry Manufacturing Plants*, World Class International Limited, Denmead, Hants.

9

JUST-IN-TIME MANUFACTURING

Alan Harrison

Just-in-Time (JIT) is the Western embodiment of an approach to manufacturing developed and perfected by the Japanese. JIT companies aim to close in on the ideal of meeting demand instantaneously with perfect quality and no waste. Thereby, competitive performance is improved continuously over time. It is unfortunate that the term implies that this approach is simply about delivering materials to the production line 'just in time'. In fact, JIT plays a much wider role in New Wave Manufacturing strategies. Many companies prefer to use alternatives like 'World Class Manufacturing' or 'Lean Production' methods. All such terms describe the quest for excellence in manufacturing along three broad fronts:

- Techniques: many so-called 'core JIT techniques' are systematically put in place to attack all sources and causes of waste. It is the combined effect of applying these techniques which makes JIT such a formidable competitive weapon.
- Everyone participates: JIT is a 'total' approach to excellence in manufacturing. This means that everyone and every process must be included. If only some of the people are involved, then only some of the problems will be solved and only some of the processes improved.
- Continuous improvement: JIT philosophy stresses the concept of 'ideals' in many areas, such as zero scrap and zero inventories. While current performance may fall far short of such high ideals at present, it is possible to get closer to them if everyone is determined to do so. Therefore, JIT has been called a 'journey with no end'.

It may already be apparent that JIT and TQM (described in Chapter 6) have many elements in common. In fact, they are very complementary strategies, and the route to excellence in manufacturing is often described as 'JIT/TQ' to emphasize the joint waste elimination/cultural aspects. Figure 9.1 shows the route to excellence as a five-stage process, and artificially separates the JIT and TQ aspects of each stage to exaggerate the differences between the

JIT EMPHASIS		TQ EMPHASIS
How do we compete? Role of JIT Initial game plan	Stage 1: BUSINESS STRATEGY ISSUES	Set the goals today to be in business tomorrow. What Deming's14 points mean in our company. Plan the transformation process
Safety Housekeeping Quality standards	Stage 2: BASIC DISCIPLINES	Work on attitudes (eg 'them and us') Work on disciplines
Process capability studies Bring process under conltrol	Stage 3: PROCESS CONTROL	Training in PS techniques All personnel involved in improvement activities. Supplier QA
Design for manufacture JIT techniques Involvement of suppliers	Stage 4: ELIMINATE WASTE	Continuous process improvement Internal customer feedback Supplier process improvement
Error proofing Automatic condition monitoring	Stage 5: ELIMINATE ERRORS AT SOURCE	Control embedded within and driven by the organization

Figure 9.1 Developing JIT/TQ
PS = Problem Solving

strategies. The sequence of the steps involved is important. First, set a game plan to identify improvement goals and the transformation process needed to achieve them. Then get the basics right – housekeeping, safety and quality standards and the attitudes which make them permanent. Then bring processes under control by involving everyone. Then extend the war on waste to every process – internal and external (suppliers, etc). Finally, aim to make it impossible for processes to be carried out any other way than the right way. Here, the TQ goal of 'control embedded within and driven by the organisation' (Foster and Whittle, 1989) is fully achievable.

The sequencing of improvement processes is fundamental. Attempts to implement quality circles or Just-in-Time supply without preparing the foundations first will at best be sub-optimal, at worst an expensive flop. The same goes for automation, including introduction of computerized systems like Materials Requirements Planning.

This chapter is in four sections. First, we review the historical development of JIT and its current relevance to the global competitive climate. Second, the JIT philosophy is defined and a brief description of supporting tools and techniques is given. These are linked to a policy for implementation in the third section, which also considers measures of performance. The fourth and final section considers management implications of Just-in-Time, including an evaluation of how Japanese management methods can be put to work in a Western setting.

THE DEVELOPMENT OF 'JUST-IN-TIME'

What we now call 'Just-in-Time' had its origins in the Toyota Production System in Japan. Introduction of the Toyota Production System was led by Taiichi Ohno, who became vice-president of manufacturing. Ohno was supported in his work by Shigeo Shingo, a consultant who among other things ran training courses in industrial engineering at Toyota from 1955 to 1981. Toyota's ability to ride the first oil shock in 1973/4, and to continue to grow, caught the imagination of other automotive companies both in Japan and elsewhere in the world. For example, Toyo–Kogyo – maker of Mazda cars – was nearly bankrupted by the effects of the first oil shock on its major market in the USA. One of the major factors in the turnaround of Toyo–Kogyo's perilous business position was the introduction of the Toyota Production System, already regarded in Japan as fundamental to manufacturing competitiveness.

Ford Motor Company bought a 25% stake in a weakened Toyo–Kogyo in 1976, and subsequently flew out thousands of executives from its American and European plants to learn about Japanese methods. In 1979 Ford UK introduced its 'After Japan' (AJ) campaign. A major feature was to invite groups of workers to form 'quality circles' – groups of workers who volunteer to meet regularly to identify and solve problems which affect their work. It had already become apparent that quality circles were a special feature of Japanese management which had a major impact on costs and quality.

Ford UK's experience was not encouraging. In April 1981 support of the trade unions was withdrawn, and it proved to be impossible to continue the development of quality circles. The effect was to stall joint improvement programmes for years. In a subsequent analysis of the situation at Ford it was concluded that the failure of quality circles was due to the attempt to use an involvement technique which was a feature of a 'very different management approach to that practised by Ford' (Guthrie, 1987).

Here was an early and very public example of a situation where Japanese management techniques did not work in the West. Ford's experience was repeated many times in companies across Europe and America. Yet an increasing number of Japanese 'transplants' (subsidiaries of Japanese companies managed by Japanese nationals) in Europe and the USA show that it *is* possible to make such methods work in the West. Schonberger (1982) describes the Kawasaki plant at Lincoln, Nebraska, in terms of 'nine hidden lessons in simplicity'. One of these lessons is that 'management technology is a highly transportable commodity'. Other examples of adapting Japanese management methods in the West are provided by Nissan (Wickens, 1987) and Yuasa Battery (Murata and Harrison, 1991).

Lean Production

A $5 million, five-year study of the automotive industry worldwide was commissioned by Massachusetts Institute of Technology in 1985. Results

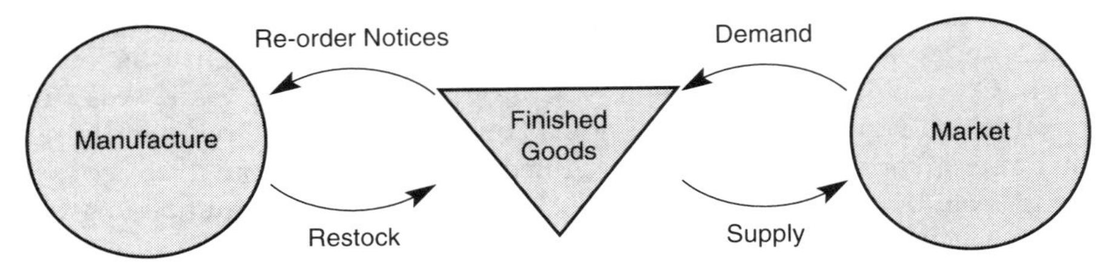

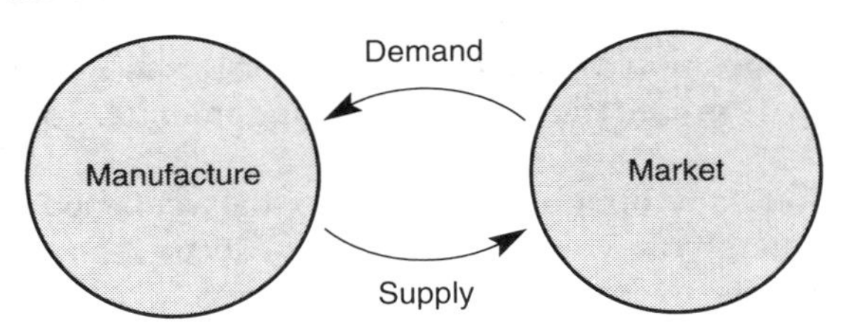

Figure 9.2 Two solutions to competitive delivery dates

were published in a series of articles and in a book called *The Machine that Changed the World* (Womack, Jones and Roos, 1990). While it was apparent that there was a clear difference in performance between 'best' and 'worst' factories in Japan, Europe and in America, 'best' plant performances were always shown by Japanese companies in Japan. Further, best performance extended not only to production but to design and to sales and distribution as well. The authors coined the term 'Lean Production' to emphasize that:

> it uses less of everything compared with mass production – half the human effort in the factory, half the manufacturing space, half the investment in tools, half the engineering hours to develop a new product in half the time. Also, it requires keeping far less than half the needed inventory on site, results in many fewer defects, and produces a greater and ever-growing variety of products.
>
> (Krafcik and McDuffie, 1989)

The implications are that product design and manufacturing/distribution can be interfaced more closely with what the market wants. As indicated in the introduction to this chapter, 'Lean Production' is synonymous with JIT.

Two possibilities for meeting customer needs, shown in Figure 9.2, are Mass Production and Just-in-Time.

Mass Production

Manufacture is decoupled from the market by means of a buffer of finished product inventory. A stable manufacturing environment is thereby created wherein large batch sizes are possible. High inventories (raw materials, Work in Progress *and* finished goods) are therefore characteristic of this approach to manufacture. Often, the very item which the customer wants is unavailable, and the stock replenishment time is relatively long. Traditional Mass Production is also characterized by work segmentation, standardization methods and firm discipline imposed through a hierarchical organization.

Just-in-Time
Manufacturing and distribution are interfaced closely with the market. Throughput times must therefore be relatively short so that demand can be met quickly. So batch sizes and finished product inventories must be relatively small. The manufacturing facility must be flexible – that is, it must be able to change what is done quickly. It must also be possible to pass on what is needed quickly from one process to the next. Order cycle times are relatively short, and the facility takes on more and more 'make-to-order' rather than 'make-to-stock' characteristics. Lean Production typically needs relatively low inventories, high reliability, flexible work teams, self-discipline and incremental improvement.

The competitive advantages of lean manufacturers who can support low-cost, high-accuracy products with short order-to-delivery lead times can be devastating in the marketplace. Lean manufacturers can use their superior design, production and delivery systems to reduce costs or increase variety, or both.

Reflective Production

Arguing that assembly line work is flawed, however you enhance it with techniques like quality circles, teamwork and JIT deliveries, Swedish authors have proposed that the Volvo approach at Uddevalla and Tuve has distinct advantages over Lean Production methods (Ellegard *et al.*, 1992). They call the approach 'Reflective Production' to emphasize that 'principles of learning rest on a man's natural way of thinking'. At Uddevalla, cars were built up in quiet rooms (no assembly lines) with magnificent views over a fjord by separate teams of assemblers. Characteristics of Reflective Production are:

- parallel material flows which enable autonomous work teams to assemble vehicles independently of each other;
- work tasks assigned to teams and individuals comprising a larger number of operations requiring longer work cycles (two hours and more).

It is proposed that Reflective Production improves human benefits like autonomy and scope of work. Figure 9.3 is the authors' view of the relationship of different approaches to production. There is currently no documentary evidence that Reflective Production yields the same economic benefits (in terms of productivity, quality and turnover of working capital) as Lean Production, so it is questionable whether the Reflective Production circle should be placed as far to the right as it is indicated. The Uddevalla plant has now been closed for car production.

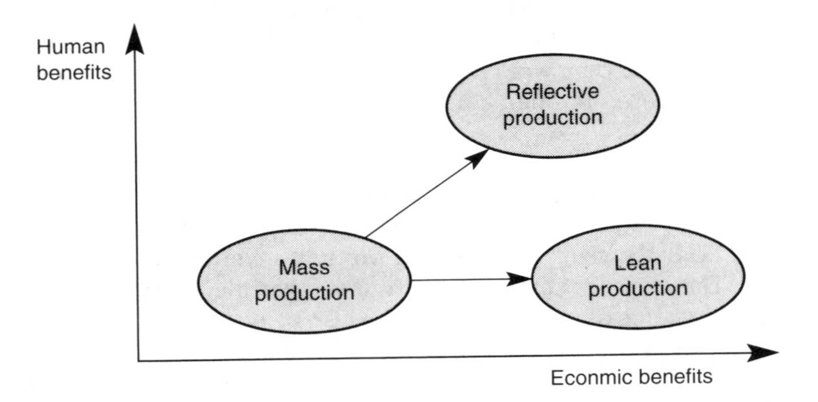

Conjectured position of different production systems for final assembly of motor vehicles.
Human benefits: high levels of autonomy, job satisfaction, growth, etc.
Economic benefits: high productivity, quality, stock turnover, etc.

Figure 9.3 Traditional Mass Production, Lean Production and Reflective Production
Note: The above are production forms each of which is based on a particular mode of exploitation of human potential and technical facilities. Efficient and humane use of technical and Human Resources results in economic benefits, e.g. high productivity, high product quality and high capital turnover, as well as human benefits, e.g. high autonomy, adequate ergonomics and broad work content. The diagram shows conjectured economic and human benefits associated with different types of production systems for final assembly of motor vehicles
Source: Ellegard et al. (1992) with permission of MCB University Press.

Other industries

So far we have concentrated on the historical development of JIT through the route taken by the automotive industry. As the world's largest manufacturing industry, this has naturally received most attention. Historically, it was the one in which Japanese ideas for improved competitiveness were first developed as a new 'production system'. But Toyota's ideas did not stop there.

During the 1970s Toyota's ideas spread rapidly to other industries in Japan. In the electrical industry, Matsushita (a much larger but less well-known company than Sony) developed its own version. Shingo thought that the Matsushita production system was better than Toyota's. Instead of using *kanban* to signal the need for more parts between separated operations. Matsushita concentrated on placing operations next to each other so that there was no need for signalling! Hence, operations were more closely co-ordinated and transport was minimized.

The philosophy of simplicity and the elimination of waste is very attractive, and has been applied in industries where there is already a high level of flow. For example, Kao Corporation manufactures soaps and detergents, haircare and cosmetic products. Such products have typically been made to stock, as in the Mass Production example shown in Figure 9.3. Kao has

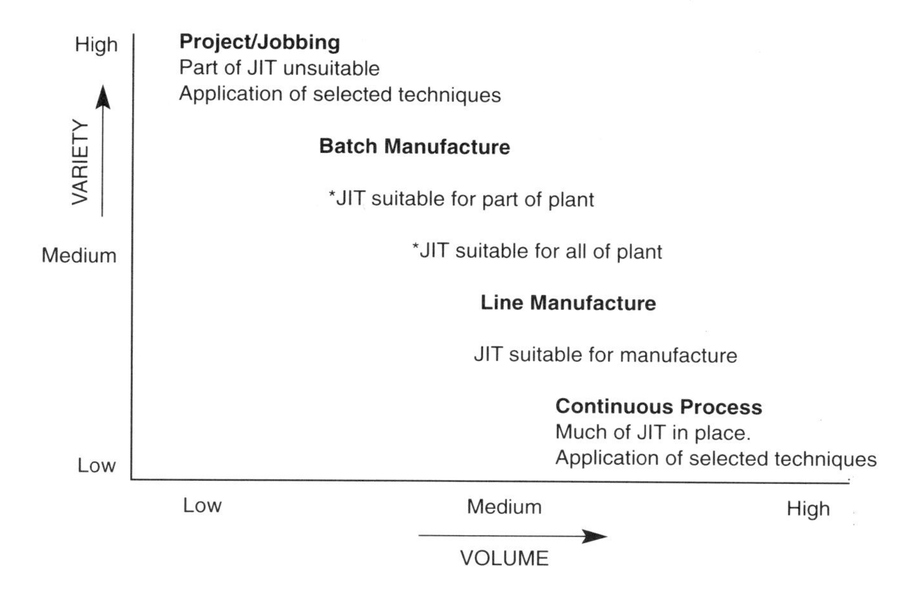

Figure 9.4 JIT and choice of process

focused on a closer integration of manufacturing and delivery systems. Retail outlets are serviced by means of daily orders to distribution centres, which aim to meet those orders within twelve hours. The distribution centres use highly automated picking methods, and load retail orders in delivery sequence. Daily sales information is provided to head office, which tracks sales by product, region and market segment by means of an integrated market intelligence system.

Figure 9.4 illustrates the suitability of JIT for a range of process choice environments. JIT tools and techniques have been particularly effective in cutting out waste from batch manufacturing environments. But jobshops can also benefit from the application of selected techniques, such as those aimed at improving flow in manufacture. Process industries have typically adopted such techniques already, but can often benefit from other techniques, such as those aimed at improving co-ordination between manufacturing and distribution mentioned in the Kao example above.

JIT PHILOSOPHY AND CORE TECHNIQUES

Just-in-Time often starts life in the role of the Manufacturing Strategy for a business. Building on the Total Quality message that 'The customer is the next process' (Ishikawa, 1985: p. 107), JIT/TQ development inevitably leads to incorporation of other business functions. Key examples are design (responsible for around 80% of manufacturing costs), sales/distribution and the supply chain. All three areas have contributed massively to the development of JIT/TQ philosophy in companies around the world.

Underpinning JIT/TQ philosophy is a set of beliefs. Two key beliefs are

- This company can become an excellent one over time. So its rate of improvement must be faster than that of its competitors.
- Excellent people make an excellent company. People are assets which grow over time. The creativity of company members is a vital input to the improvement process. Company members must themselves grow over time to be capable of handling the more exacting jobs which result.

Such beliefs are often incorporated into a company's statement of mission, values and guiding principles. In turn, such beliefs are supported by three basic elements of JIT/TQ philosophy. These are: (1) the elimination of waste; (2) Total Quality; and (3) people preparation.

1. The elimination of waste

Seven wastes have been found to apply to many different situations:

Overproduction
The greatest source of waste often results from producing more than is needed by the next process. It is 'safer' to work ahead of the next process so that there is a comfortable buffer-stock to cushion potential problems. On the other hand, overproduction leads to problems like double handling, longer lead times, extra Work in Progress, and lack of responsibility for quality.

Waiting time
While machine efficiency and labour efficiency have been two popular measures in UK industry, real waiting time is often not visible. For example, materials may only be worked on for 9% of the time they spend in the factory. Much of the remaining 91% of time is spent in storage or waiting in queues of Work in Progress (or rather, work not in progress!).

Transport
This is another activity which does not add value to the product. Powered conveyors can become a form of automated waste. For example, Raleigh Industries – who manufacture bicycles in Nottingham – found that a cycle frame travelled a distance of six and a half miles through the factory! This has subsequently been considerably reduced. Placing processes closer and closer together helps to cut out this source of waste.

Process
The process itself may become a form of waste. A machine which runs faster than the next process creates waste. Operations like deburring and adjustment do not add value. A process which is not under statistical control, or which is incapable of meeting specified tolerances, is too variable, and creates defects.

Inventory
The presence of inventory anywhere in the production system implies that there are unresolved problems. To the JIT/TQ company, inventory is an evil. So the presence of inventory in the stores (raw material or finished product), as 'work not in progress', as buffers and large batch sizes is always a symptom that there is scope for improvement.

Motion
People can often look busy when the work they are doing does not add value. For example, searching for missing tools or collecting another card from the supervisor's office are wasted time. Other improvements can be made to the job by for example reducing bending and reaching movements.

Defective goods
The visible part of defective goods (scrap notes and concessions) is often dwarfed by the invisible parts like errors and delays to flow of production, excess inventories like 'scrap allowances' and customer complaints.

2. Total Quality

Some points need to be emphasized here in addition to those already made in Chapter 6. 'Total' refers to everyone and every process. 'Quality' refers to everything the company does from providing goods and services which meet customer requirements to the way that company members interact, to what the company expects from its suppliers. Total Quality is fundamental to becoming an excellent company and provides the necessary development of the organization which enables company members to join in the war on waste.

Figure 9.5 summarizes key aspects of TQ which impact on the JIT philosophy. In Japan, a foundation of TQ was developed in many companies under the auspices of Drs Deming and Juran before the Toyota Production System – or other new production systems – were developed. It is difficult or even impossible to attempt to introduce involvement techniques for waste elimination in a traditional organization culture – witness the Ford experience described in the previous section.

3. People preparation

'Hiring the whole person' means hiring someone not only to do a job of work, but also that person's brain, senses and potential for further development. Figure 9.6 summarizes the key aspects of people preparation:

- Discipline: is the critical essence of a manufacturing company. Without it, there will be a lack of quality and safety standards. Discipline is a shared set of values, rather than an imposed set of rules.

Figure 9.5 Total Quality: JIT basic element 2

- Flexibility: refers to the expansion of people's capabilities by means of consistent, long-term programmes of training in new skills.
- Equality: is the removal of unfair personnel policies which create divisions between groups of company members, such as different pay structures and staff car parks.
- Autonomy: results in the delegation of authority for running the work area to company members who carry out the work itself. Examples of autonomy are authority to stop the line, routine material control and problem-solving activities.
- Quality of Working Life: refers to measures which set out to improve the sense of involvement of company members, their security of employment, and their sense of enjoyment of working life.
- Creativity: harnesses the natural curiosity of company members to make improvements which affect the work they do. While this may seem to be opposed to the concept of discipline, if the aim of the job is made clear, company members can be given discretion as to how it is carried out.

The three basic elements of JIT/TQ philosophy can be pictured as a set of overlapping virtuous circles, as in Figure 9.7. A balanced approach is needed: each of the three areas must be addressed in planning for an excellent company. The whole system is stronger than its parts, so it is essential to take a holistic view.

Figure 9.6 People preparation: JIT basic element 3

Figure 9.7 Basic elements of the JIT/TQ philosophy

Core JIT techniques

In the engine room of a JIT/TQ company is a set of tools and techniques which are for use in the war on waste. There are many of them, and they follow on naturally from the JIT/TQ philosophy. Figure 9.8 shows a selection of such techniques arranged in a circle. Those on the right-hand side of the figure (from design to set-up reduction) have been referred to as 'JIT1' (Bicheno, 1991). They are techniques which help to prepare the facility for reaction speed, low cost, short lead time and high-quality manufacture. Such techniques are universal best practice. Techniques on the left-hand side of the figure (from flow scheduling to co-makership) are referred to as 'JIT2'. They are techniques which carry forward the war on waste. Only a brief summary is possible here: for further details, see Harrison (1992).

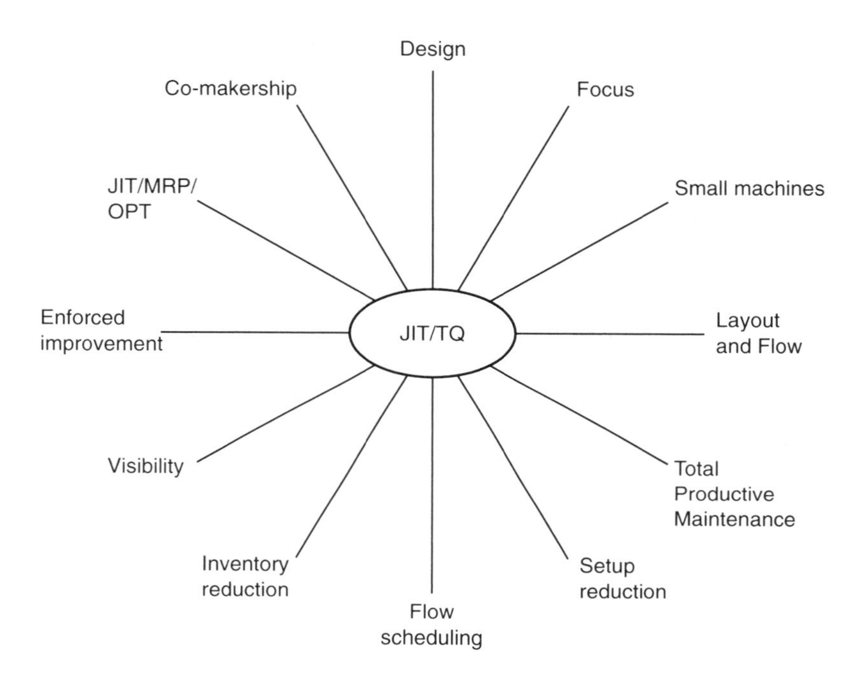

Figure 9.8 Selected JIT techniques

JIT1

Design

Design improvements can halve the product cost through quantum reductions in the number of components and sub-assemblies, and better use of materials and processing techniques (Corbett, 1986). Multidisciplinary teams help to reduce costs and time to market, as well as improve quality because manufacturing processes are considered alongside design. 'Design for Manufacture' is the umbrella term used to cover the tools and techniques which make this possible. The aim is to 'focus design team effort on the cost effective use of parts and processes to produce on time, high quality products that meet customer and business requirements' (Miles, 1990: p. 1). See also Chapter 4.

Focus

The concept of focus is that the manufacturing task has been limited to a simple, consistent and achievable set of goals. The focused factory will always outperform the unfocused factory because it does a few things extremely well. Often, this is achieved by means of the 'plant-within-plant' approach, whereby the factory is divided up into a number of product groupings, each 'sub-plant' with its own set of plans and policies. Product focus is a major feature of the best JIT companies (Safayeni *et al.*, 1991). A development of this concept is Cellular Manufacture, discussed in Chapter 10.

Small machines

Several small machines, perhaps permanently set up, are used in preference to a single, large machine. Large machines are more prone to creating bottleneck problems, lengthy maintenance and repair, and to large inventory needs. For example, benchtop flow solder machines can help to reduce lead times and inventory by avoiding the need to transport WIP to and from a large, central facility. The trade-off between reduced WIP and increased investment in machinery is a legitimate one! Small machines help to reinforce the concept of product focus.

Lay-out and flow

Shingo (1989) saw a production system as a network of operations and processes:

- Processes are the chain of events by which raw materials are converted into products.
- Operations are the chain of events by which workers and machines work on parts on products.

Figure 9.9 is my version of Shingo's process/operations matrix. Follow a batch of parts (shafts or bushings) through the processes. Starting at the base of the figure, a batch of shafts must first wait in stores (triangle), then it must be transported for processing (arrow). Then it is delayed while waiting for an operation (D). Then the operation takes place (O), after which the batch of parts must wait again for inspection (D). After inspection (square), they are stored again (triangle) before shipment. Shingo teaches us that his matrix is the correct way to look at production. First, aim for efficiency of processes by continuously attacking the waste. This includes action like moving operations closer and closer together to avoid the waste of transport. At each stage of improvement, operations are fitted in around the processes. In the West we have been preoccupied with efficiency of operations first, leading to concern for labour and machine efficiency at the expense of lead times and Work in Progress.

Total Productive Maintenance (TPM)

Unplanned breakdowns are a source of variability leading to waste. TPM aims for perfect equipment reliability by involving everyone in the quest for

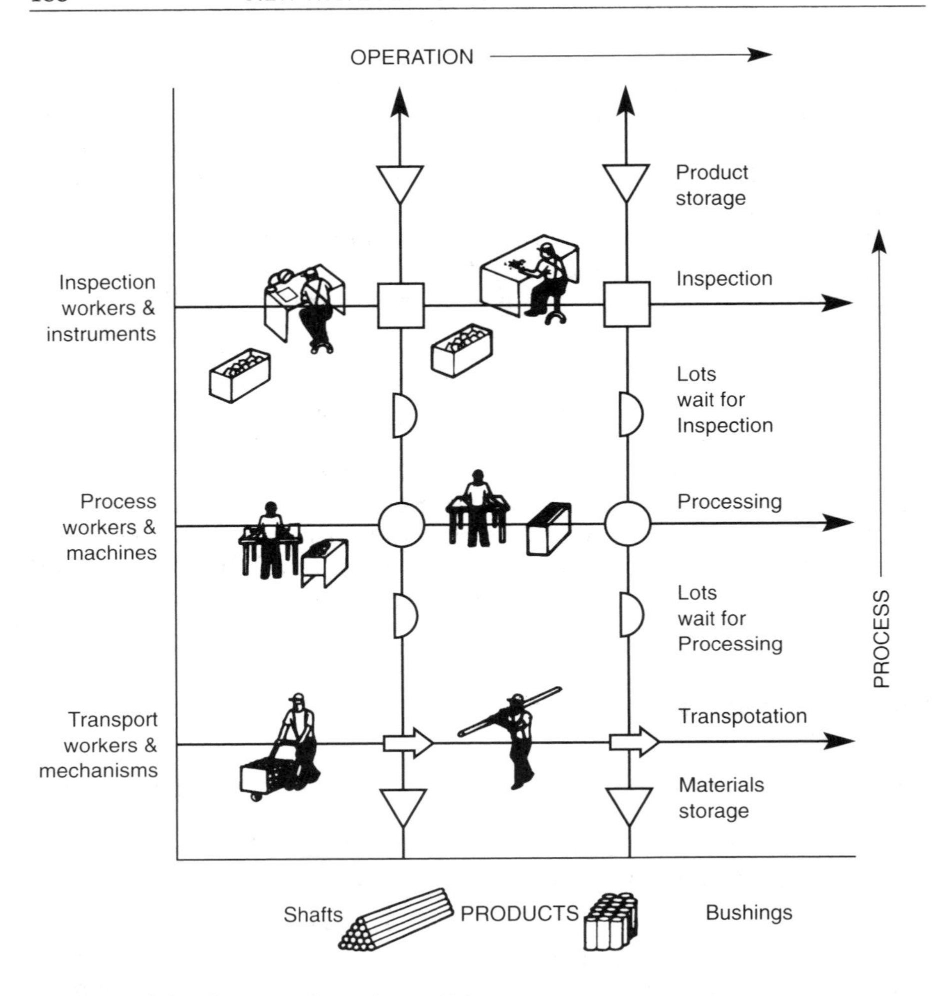

Figure 9.9 Shingo's process/operation matrix
(Source: *A Study of the Toyota Production System from an Industrial Engineering Viewpoint*, Shigeo Shingo, English retranslation copyright © 1989 by Productivity Press, Inc., PO Box 3007, Cambridge, MA 02141, USA)

improvements. The attitude by production operators that 'maintenance is not my problem' is a killer! Instead, production operators can help to prevent breakdowns by

- machine cleaning, which is also a form of inspection;
- accurate machine operation;
- using their five senses to detect impending problems;
- carrying out routine maintenance, such as lubrication.

This allows the maintenance specialists to carry out more improvement activities. Time must be allocated for maintenance activities, and undercapacity scheduling helps to provide the necessary windows.

Set-up Reduction (SUR)

Improving reaction speed in a production system, so that it can interface more closely with the market, requires that batch sizes and hence WIP are reduced. Systems for generating fixed batch sizes – such as economic order quantities – should be questioned, especially for the most expensive (Class 'A' and Class 'B') parts. The only way that smaller batch sizes can be achieved without reducing the capacity of the system is by reducing set-up times. Set-up time is defined as the time taken from the last good piece from the preceding batch to the first good piece from the succeeding batch. Often, set-ups can be reduced by simple and inexpensive means, such as cutting out search time for the next tool, and preparing dedicated tool change trollies with the necessary kit. The SUR process is typically an ideal opportunity for a TQ approach using multidisciplinary teams which are empowered to implement their ideas.

JIT2

Flow scheduling

Operations are linked together by means of 'invisible conveyors', whereby the succeeding operation signals to the preceding operation that more parts are needed. This process is called 'pull scheduling', and the signal is often referred to by means of its Japanese name '*kanban*'. In this way, the aim over time is to synchronize all operations throughout the factory, so that they are co-ordinated to a 'drum beat' from the Master Production Schedule. Accurate inventory timing (neither too early, not too late) is the aim. *Kanban* is a simple-to-operate, visible method for delegating routeing material control activities to the shopfloor. Rules for operation are simple but strict. They are shown in Figure 9.10.

1. Each container must have a *kanban* card, indicating part number and description, user and maker locations, and quantity
2. The parts are always pulled by the succeeding process (the customer or user)
3. No parts are started without a *kanban* card
4. All containers contain exactly their stated number of parts
5. No defective parts may be sent to the succeeding process
6. The make (supplier section) can only produce enough parts to make up what has been withdrawn
7. The number of *kanbans* should be reduced

Figure 9.10 *Kanban* operation rules

Inventory reduction

Inventory in a manufacturing system can be considered under three headings:

- Before manufacture: 'invisible inventories' which result from delays like order processing and preparation of designs.

- During manufacture: high inventories and long lead times caused by large batch sizes and buffer-stocks.
- After manufacture: inventories caused by delays in delivery to customer and in receiving payment.

Inventory reduction is about systematically attacking the causes of all such forms of waste, for example by means of batch size reduction, inventory accuracy disciplines and simplified bills of material.

Visibility

Control is greatly enhanced when standards, needs and problems can be seen. So a JIT factory is characterized by simple, visible control systems like *kanban* boxes or square or tokens, and by cockpit-style checklists for set-ups and maintenance. The team's latest quality projects are visible too, as evidenced by fishbone diagrams in the team's meeting room. Visibility is also enhanced by the avoidance of partitions in offices and pallet stacks in the factory that are more than five feet high. Because so many aspects of performance are now current and visible, 'management by walking about' – directed at helping to solve problems – is more effective than supervision based on last month's figures!

Enforced improvement

The famous analogy of the ship and the rocks is shown in Figure 9.11. Problems such as late and defective materials and machine downtime (the rocks) have been covered with a sea of inventory so that the boat can float. Enforced improvement aims deliberately to confront the problems, and by finding solutions to the basic causes of the problems allow the water level to be reduced. Taking out the inventory before the problems have been solved is a more risky action to take. But this may be done on an experimental basis by the team to find out where problems are located in the current system.

JIT/MRP/OPT

JIT philosophy has much to contribute to the effectiveness of computerized systems such as Manufacturing Resource Planning (MRPII, discussed in Chapter 8). These include:

- JIT1 preparation of basic disciplines is of value to all manufacturing planning and control systems.
- The number of transactions on MRP can be reduced (often by as much as 70%) by JIT simplification.
- Fixed scrap rates, batch sizes and lead times are anathema to the continuous improvement process.
- MRP is poor at control, good at planning in most applications. JIT pull scheduling is good at control, but has little to offer the planning process. There are possible synergies here.

- The capacity of bottleneck operations on OPT (Optimized Production Technology) can be improved by Set-up Reduction.

The key point is to develop an integrated approach which is best suited to a given situation. Today, there is an increased number of opportunities for innovation and excellence.

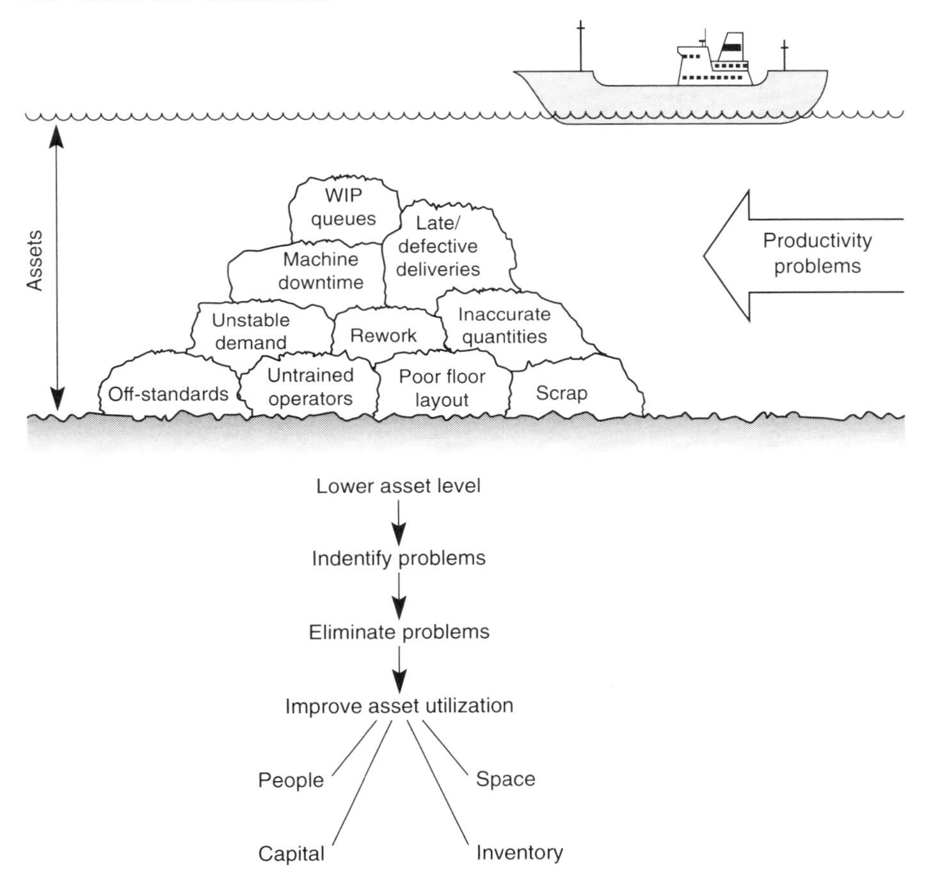

Figure 9.11 Enforced problem solving

Co-makership

The supply chain essentially consists of a series of customer–supplier links. Ideally, the links recognize their dependence on each other and their contribution to the competitiveness of the supply chain as a whole. Such recognition can lead to joint development programmes, partnership agreements, integrated systems and even to mutual investment between major links. Responsibility for 'conducting the orchestra' is that of the dominant supply chain member, such as major retailers and vehicle assemblers. Carriers have a major role in maximizing the efficiency of material movement timing, and in cutting out waste. This is a far cry from the situation where business is placed competitively on an annual basis to the lowest of three bids!

Conclusion

While it has been possible to list only some highlights of selected JIT techniques, it should be possible to see that a complex, overlapping series of actions can be undertaken in pursuit of the overall JIT philosophy. The specific actions which are taken will depend on the company and its existing capabilities in relation to its customers and competitors. But a rich and powerful strategy can emerge. A virtuous cycle can develop – it is not necessary to have JIT1 in place before JIT2 techniques are started. For example, Set-up Reduction may help to reduce batch sizes and buffer-stocks (JIT2). This in turn may help to improve lay-out (JIT1), and hence to improve opportunities for visibility (JIT2). At each advance, keep asking yourself: what opportunities does this improvement open up? It is this relentless search for excellence which makes JIT such a formidable competitive weapon.

IMPLEMENTING JIT

In a comparison of JIT applications in the USA and the UK (Billesbach *et al.*, 1991), we concluded that

- There was a significantly higher level of JIT activity in the USA than in the UK.
- Training in JIT was focused on higher and middle management in the UK, but was spread across all levels in US companies. Significantly less training was provided overall in UK companies.
- US companies placed a greater strategic emphasis on JIT. UK companies suffered from a significant lack of knowledge and education in a number of JIT activities.

The main points from the survey are shown in Figure 9.12. These results, and our experience in visiting plants in both countries, led us to propose two types of JIT application:

Percentage of companies	USA	UK
Using JIT	90	75
Using consultants	77	19
With operator involvement in problem-solving	69	48
Using operator inspection	88	74
Giving line stop authority	68	39
With operator maintenance	40	23
With paid suggestion schemes	38	47
Average training (hours/year)	20–40	<20
No. of responses	68	64

Figure 9.12 JIT USA/UK survey

- strategic JIT: where the company takes a long-term view of JIT implementation, and acknowledges its impact on the total business;
- transitory JIT: where the company sees JIT as another short-term, temporary management technique which can yield some inventory or labour cost savings.

An example of transitory JIT thinking is provided by the following view: 'JIT stands for just-in-time. That means that our suppliers are being told to deliver that way. This greatly helps to reduce our stocks.' No wonder that many suppliers look on JIT as standing for 'Japanese-Inspired Terror', when the assemblers are trying to grab all the advantages themselves! The reality of such a situation is that costs and stocks get passed down the supply chain, and that no net improvement is achieved. The assemblers get their own back in the form of higher piece part prices.

Five key questions

It has been emphasized repeatedly that JIT/TQ must form part of the business strategy. This implies that the vision of an excellent company must be specified, together with the role of JIT/TQ in achieving this vision. Voss and Clutterbuck (1989) proposes that the top team asks itself five key questions before starting on the JIT journey.

Will JIT improve our business performance?
JIT deliverables such as reduced costs and time to market need to be considered in relation to their relevance to current and future business needs. If a company decides not to adopt JIT, then it should consider the potential threat from competitors who will, or who already have.

How suitable is JIT for our manufacturing environment?
Thinking that JIT is limited to manufacturing environments where there are high, stable volumes is mistaken. First, it takes only a narrow view of 'JIT' (flow scheduling). Second, it ignores the opportunities to take more and more parts out of complex, error-prone production control systems into the relative simplicity of pull scheduling by product and process simplification. JIT/TQ philosophy and techniques are applicable across a wide range of situations (see Figure 9.4).

Should we invest in JIT, new process technology (for example CIM), or both?
JIT1 techniques help to develop competence and simplification in the way things are done. JIT2 techniques help to squeeze out waste and thereby to integrate processes. It is proposed that these are essential precursors to automation. The major risks of skipping such necessary preparation are that waste will itself be automated and that the necessary learning will not have developed.

How should we implement JIT?
A typical organization for implementing JIT would be as follows: the steering committee handles overall co-ordination, the JIT champion – preferably full time – handles day-to-day facilitation, and implementation teams work on specific projects on a part-time basis. Both steering committee and implementation teams have cross-functional membership. Pilot projects normally precede full-scale implementation.

What fundamental changes do we have to make to become a JIT company?
Fundamental changes include the transformation towards a Total Quality culture, with its accompanying demands on leadership, growth of company members and new ways of defining and measuring performance.

Satisfactory answers to these questions need to be made as part of the meticulous planning process needed to achieve the transformation to an excellent company.

New measures of performance

While the 'bottom line' will always be an important business consideration, financial measures have typically been used to 'control' internal processes when they are incapable of so doing. The bottom line is the net effect of the performances of all such internal processes. It is necessary to understand how internal processes operate, and how they should contribute to overall business objectives. Often, the controls which are needed are not financial in nature. Examples of new measures of performance include:

- P:D ratio: this is the ratio between the actual product lead time (P) and that expected by customer (D).
- Flow factor: the ratio of the manufacturing lead time for a given component or series of processes, and the sum of the operations times. In Shingo's terms, it is the ratio between the total process time and the sum of the operations times (Figure 9.9).
- Flow distance: records the actual distance travelled by a product during the course of manufacture.
- Defect ratio: the number of defects per number of good parts. Initially, the ratio of defects to parts might be one per thousand, but this could be increased to one per million or more as improvement is made.
- Space requirements: records the number of square feet taken up by a given product or component.
- Maintenance measures: record issues like mean response time to breakdowns, mean time to repair (MTTR), and total uptime per week.

Selection of appropriate measures of performance is vital. They therefore reflect managerial priorities. What gets measured is what gets done.

Sequence of implementation

Starting off on the JIT journey is not like going round a supermarket! You cannot pick up the goods in any order you please. Attempting to short-cut the sequence is likely to lead to transitory JIT. JIT implementation should be addressed in planned stages, as shown in Figure 9.1. Expanding on the summary given at the start of the chapter, the suggested stages are:

Stage 1: strategy issues
This examines the fit between JIT and the business strategy, and addresses the five key questions listed above. If the answers are positive, then a plan needs to be developed for how the organization will go about becoming a JIT company.

Stage 2: basic disciplines
A firm foundation must be made which ensures that good habits are developed and basic disciplines shared. Cleanliness and tidiness, including workplace organization, are fundamental. Shared values in safety and quality must also be built in here. Often it is necessary to confront bad habits that are the legacy of traditional attitudes, such as flouting safety standards.

Stage 3: process control
Once basic disciplines have been established, the next logical step is to use the new learning to bring processes under control. JIT1 techniques are aimed at this area. So are process capability studies and SPC (Statistical Process Control). The mission is low-cost, high-quality, short lead time manufacture.

Stage 4: eliminate waste
Once processes are coming under control, then the war on waste can start in earnest. JIT2 techniques are aimed at this area. Development of customer–supplier relationships within and outside the organization naturally leads to JIT concepts being extended to design, distribution and the supply chain.

Stage 5: eliminate errors at source
The ultimate aim is to produce instantaneously with perfect quality and no waste. So use of 'poka yoke', or error-proof devices, comes into play so that it becomes impossible to produce defects. Automatic condition monitoring of machines helps to signal process or equipment problems immediately. As noted in the previous section, there will be a certain amount of dynamism between the stages, for example between JIT1 and JIT2 techniques. But the need for evolution towards an excellent company must be understood.

MANAGEMENT IMPLICATIONS

During the development of JIT/TQ, the whole management approach needs to be thought out again from basics. The change can be pictured as in Figure

9.13. On the left we have the traditional organization pyramid, with a top-down command structure. Many of the tasks, like production planning and control, are non-value-added operations. Overhead rates are relatively high. On the right, there is the organization structure for a JIT/TQ company. The aim is to support the coalface, not to supervise it! This is achieved by relatively few management layers, and greater empowerment of company members who are at the coalface. The role of management is to lead by example towards the goals of excellence, and to help company members in the process of achieving them. Rather than 'I'm behind you', it's 'here's the way'!

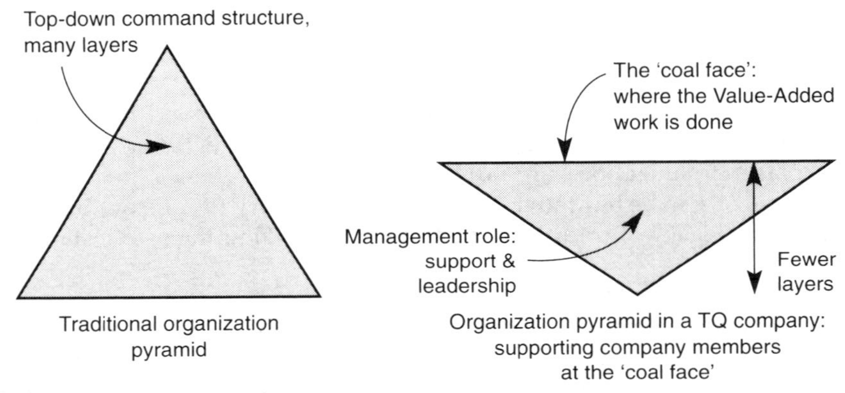

Figure 9.13 Traditional and Total Quality views of the organization pyramid

Murata and Harrison (1991) propose that there are two fundamental factors for success in a business:

- ensuring that the company advances in the right direction for the future (that is, that winning strategies are formulated and pursued);
- increasing the output of all company members.

The first of these is clearly a management responsibility and needs to reflect the business aims of the company. Business aims need to be clearly understood by all company members. It is the second factor which seems to have given Western management the greater difficulty. Increasing output of company members is not simply about getting them to work harder and faster. Improving people's capabilities is more important than increasing their output of work. Improved capabilities mean that company members are better equipped to solve problems which affect their work and to find improved methods. The motivational effects are of a different order of magnitude. Apart from the beneficial results of improvement in their own right, another powerful motivator comes into play – the Hawthorne effect (motivation comes from recognition). Schonberger (1986) describes the Hawthorne studies at Western Electric as 'the second major event in the history of manufacturing management', the first being scientific management and the third

JIT/TQ. But achieving this energized state is not an easy task. It requires great determination over time by the top team.

Making Japanese management methods work

When Kazuo Murata, Japanese managing director of Yuasa Battery of Ebbw Vale in south Wales, was taken to Cardiff Arms Park, he was greatly impressed by the passion and unity of the crowd and by the great energy in life shown by the players as they created winning moves under pressure from the opposition and under strict rules. 'This is the atmosphere that industry wants to create,' he thought. While Japanese workers typically have strong commitment to work, they have less interest in life outside work. British workers tend to react the opposite way. Ask someone who loves fell walking to go up a steep hill with a heavy backpack and he'll love it, even if it's raining! Ask the same person to carry a few lead ingots to the foundry and he'll pull a long face.

From such simple reasoning, we developed a theory for transforming Western attitudes. A version of our transformation diagram is shown in Figure 9.14. That Japanese workforces are so good at manufacturing is to a

Successful Japanese Way	Remove Japanese Characteristics	Essential Points	Add Western Characteristics	Successful Western Way
Traditional enjoyment of teamwork	Tradition from feudal past Rebuilding after WW II	Teamwork Discipline	Love of sport Rules of sport apply to all	Eliminate 'them & us'* Shared values at work +
Loyalty to the Company	Group loyalty	Loyalty	Individualism	Lose dislike of work * Personal development +
Flexibility at work	Rebuilding after WW II	Flexibility	Hobbies, sports, DIY	Eliminate demarcation * Promote indiv. interest +
Work is respectable	Society view	Value of working life	Enjoyment of life outside work Time consciousness	Lose dislike of work * Enjoyment of working life + * unlearning + learning
QC Circle motivation	Group behaviour	Creativity	Hobbies, sports Sense of humour Creative flair	Confront 'improvement is none of my business' * Encourage creativity +

Suggested Development Sequence ↓

Figure 9.14 Transforming Japanese methods into Western success.
Key: * Unlearning; + learning

large extent because JIT/TQ values fit very comfortably with Japanese cultural characteristics. So features of the 'successful Japanese way' in the left-hand column are ideally suited to World Class Manufacturing. In this sense, I have come to believe less and less in Schonberger's (1982) fourth lesson that 'culture is no obstacle'. But if you remove the Japanese characteristics in the next column, then you are left with a set of 'essential points' for excellence in manufacture which would be independent of culture.

The next stage in the argument is to add Western cultural characteristics which are appropriate to individual essential points. There are plenty to chose from. Western values of sport and games naturally complement the essential points of teamwork and discipline. Our hypothesis is that it is possible to transfer such values into the world of work. Such a transfer will have two main effects:

- Learning: to develop shared values at work in the same way as shared values are already held in sport and games.
- Unlearning: to confront and eradicate traditional Western bad habits such as 'them and us' divisions.

Each of the other essential points needs to be addressed in such a way, with sensitivity as to how best to develop the enhanced state of motivation which is a feature of world-class companies. It is suggested that the five areas in Figure 9.14 form a development sequence. The challenge is for Western companies to achieve the essential points better than worldwide competitors.

The industrial relations impact of JIT/TQ

A standard view in the field of industrial relations is that JIT is risky in the British context because industrial conflict makes employers especially vulnerable. The dispute at Lucas Electrical in October 1986, when the imposition of an overtime ban stopped production at Austin Rover, is said to 'highlight the susceptibility of JIT to even low-cost forms of industrial action' (Turnbull, 1988).

The inevitable response to guard against such susceptibility is to hold buffer-stocks. But cushioning weaknesses in the supply chain (Turnbull calls them 'choke points') by holding buffer-stocks is anathema to JIT. There is no half-way house in this argument. Either you embrace the Mass Production philosophy and the increase in inventory it brings, or you embrace the JIT philosophy and set out to eliminate the weaknesses themselves. If you continue to hold stocks to buffer forms of industrial action, then you will lose out to companies who do not (or to subsidiaries where it is unnecessary). When some Ford national trades union officials claimed that the UK could stop other Ford plants in continental Europe by halting engine production at the new plant at Bridgend, Ford's reaction was to divert phase two of the development from Bridgend to Cologne.

In my experience, the implementation of Japanese methods has never been seriously impeded by such considerations. Far more failures have resulted from management weaknesses which have resulted in various forms of transitory JIT.

Implications for human relations plans and policies

It would not be workable to aim for development towards a JIT/TQ company with traditional UK personnel policies in place. For a start, how can you expect people to increase their output when all that will happen is that they work themselves or a mate out of a job? It must be safe for people to participate. Deming (1986) describes this process as 'driving fear out of an organisation', an ideal which is easier to talk about than to achieve! While Rover Group has recently implemented a 'New Deal' which includes employment security for all grades up to junior managers, this is only part of the issue. New policies cannot simply be overlaid on to existing management practices, or little improvement will result. Nevertheless, Rover's move – particularly in the currently depressed economic climate – is a brave one.

Then there is the question of the understanding of work. How do we get people to change their way of thinking from the 'them and us' syndrome? Figure 9.14 proposes processes of learning and unlearning. While people's intrinsic psychological needs cannot be changed, it is possible to change their perception of work. For example, there is a world of difference between the management role of leading and helping, and that of controlling and commanding. As long as manual work is regarded as low status, the latter role will prevail. A key part of the transformation process is to recognize that the coalface (see Figure 9.13) is where value is added. To put it more simply, we are all in the same boat.

Such considerations lead me to propose the following examples of Human Resource policies to support JIT/TQ:

Selection

Yuasa Battery considered that the selection process is influenced largely by whether you consider a new recruit to be a component or an asset. If the applicant is to be a component, then you are interested in his or her current ability. But if he or she is to be an asset, then you are interested much more in the person's future potential. So an example of a 'component' would be a welder for a welder's job. An 'asset' for the same vacancy would be someone who has relevant interests in life such as car maintenance, is keen on self-improvement, and who has the potential to become team leader. Such a person would be able to pick up welding skills as the start of a process of personal development, which can be further supported by a policy of 'promote from within'.

Training and development

In a JIT/TQ company there is a much greater emphasis on empowerment to take decisions at the coalface. A long-term process of developing capabilities is needed. In part, this is aimed at cross-training so that people can transfer from areas of low demand to areas of high demand. But empowerment means developing capabilities in other areas which impact on the job – tasks which have previously been carried out by specialists like inspectors and production control. An example of this approach was the training strategy used by Cummins Engine at Daventry for Set-up Reduction on the block line. Here, two operators from the subject machine were trained with the process engineer, maintenance operator and tool control representative. 'Action teams' were based on this group to study the set-up, and to identify and implement improvements. While staff experts could no doubt have made changes, this approach put the shop experts in the front-line and ensured a much greater degree of commitment to the outcome. There is so much to be done in this area alone. Rexel Engineering, makers of office equipment, described it as 'Progress Equals Our People's Learning Experience.'

Payment systems

Simple payment structures with few grades and common terms and conditions help to promote the feeling that 'we are all in the same boat'. Sharing the agony helps as well – for example, Hewlett Packard cut pay across the board by 10% in 1985. On the other hand, large increases to directors accompanied by below-inflation increases to more junior levels cannot help. Direct incentive schemes which reward individual effort also fail to promote flexibility and a sense of common purpose. When Albion Pressed Metal, a Cannock-based manufacturer of pressed parts to the automotive industry, planned the change to a JIT/TQ environment, the piecework scheme was a matter for concern. The scheme encouraged a high level of productivity, and it was felt that this could be lost on moving to a group-based incentive after changing to a cellular lay-out. But piecework, with its emphasis on high output whether or not the next process needs it, was clearly inappropriate to the new ways of working. The decision was to abandon piecework: productivity did not reduce. Nissan UK has introduced staff conditions for all. While new thinking on payment schemes helps, the more radical change is in attitude.

Supervision

Traditionally, the supervisor role has been filled by people who are underpaid and under-empowered. Yet the role is a key one for assuring good industrial relations and high productivity and quality at the workplace. Procter and Gamble (manufacturers of soaps, detergents and toilet goods) gave the first-line supervisor (called a 'line manager') full responsibility for production (including liaison with related departments), safety, quality, costs, relations with department employees and with other managers. The posts were staffed fifty-fifty by graduates and by internal promotions from chargehands. In a JIT/TQ company, the role of supervisor also demands personal qualities of energy and leadership. The role is mostly about looking ahead and managing the improvement process. At SP Tyres in Washington, supervisors have developed their own training programmes for people on their sections. This ensures relevance, and that supervision is seen to be committed.

Appraisal
Deming (1986) counsels against the use of performance appraisal and merit rating schemes because they focus on the result, not on the process (leadership to help people). He goes on to say that such schemes encourage people to focus on short-term results, and they increase the variability of people's performance. One company I know well took this literally, and discontinued their management appraisal scheme as part of a Deming-inspired TQM programme. The scheme was replaced by a less specific 'process improvement plan', which managers never properly understood. The result was a watery environment in which management performance was not monitored or reviewed. Nissan UK have developed their own version of appraisal whereby the objectivity of the rating system has been given considerable attention. The system is also applied to all employees (not just to management, and is linked to annual merit reviews (Wickens, 1987).

Communications
Communications take on a new set of meanings under JIT/TQ. The Macintosh division of Apple Computers described the management task as 'heavy duty communications'. If company members are not informed, then how can they do the right things? Knowledge and information need to be shared as fully as possible. At Yuasa Battery, communications were carried out at many levels. One of these was that the MD and senior managers made a 'helping' plant tour for one hour every morning. This helped to break down barriers between staff and shopfloor and promote direct communications. Another level was a daily staff meeting which was held in standing position so that people could form small sub-groups to discuss specific problems in parallel. Visibility of problems and data was promoted by display rooms (performance records, competitor products, etc.), by the open-plan general office (which included the MD's desk as well as other senior managers) and by company mottoes placed around the factory and offices.

CONCLUSION

The 'Japanization' of British industry has received much attention in recent years. Oliver and Wilkinson (1992) never really got round to ruling on whether implementing Japanese techniques would either:

- restore our competitiveness by ensuring business survival and success

or

- intensify work and leave a greater concentration of power in the hands of capital.

While the answer may be uncertain, what does matter is that Japanese methods have changed our thinking irrevocably. What is needed is a more imaginative approach to excellence in manufacturing. Today, there is a greatly enriched range of tools and techniques. An ever greater onus is placed on managers to create new winning strategies for the business which they lead. Human Resource practices which achieve commitment to such strategies are fundamental to success.

REFERENCES

Bicheno, J. (1991) *Implementing Just-in-Time*, IFS, Bedford.

Billesbach, T., Harrison, A. and Croom-Morgan, S. (1991) Just-in-Time: a US–UK survey, *International Journal of Operations and Production Management*, Vol. II, no. 10, pp. 44–57.

Corbett, J. (1986) Design for economic manufacture, *Annals of CIRP*, Vol. 35, no. 1, p. 93.

Deming, W.E. (1986) *Out of the Crisis*, MIT Center for Advanced Engineering Study, Cambridge, MA.

Ellegard, K., Jonsson, D., Engstrom, T., Johansson, M., Medbo, L. and Johansson, B. (1992) Reflective production in the final assembly of motor vehicles – an emerging Swedish challenge, *International Journal of Operations and Product Management*, Vol. 12, nos 7/8, pp. 117–33.

Foster, M. and Whittle, S. (1989) The total quality management maze, *The TQC Magazine*, May, pp. 143–48.

Guthrie, G. (1987) AJ and beyond, *Production Engineer*, May, pp. 29–31.

Harrison, A. (1992) *Just-in-Time Manufacturing in Perspective*, Prentice Hall, Hemel Hempstead.

Ishikawa, K. (1985) (trans, by Lu, D.) *What is Total Quality Control: The Japanese Way*, Prentice Hall, Englewood Cliffs, NJ.

Krafcik, J. F. and McDuffie, J. P. (1989) *Explaining High Performance Manufacturing: The International Automotive Assembly Plant Study*, IMVP, MIT Press, MA.

Miles, B. (1990) *Design for Manufacture and Assembly*, Lucas Engineering and Systems, Solihull.

Murata, K. and Harrison, A. (1991) *How to Make Japanese Management Methods Work in the West*, Gower, Aldershot.

Oliver, N. and Wilkinson, B. (1992) *The Japanisation of British Industry*, Basil Blackwell, Oxford.

Safayeni, F., Purdy, L., van Engelen, R. and Pal, S. (1991) Difficulties of JIT implementation: a classification scheme, *International Journal of Operations and Production Management*, Vol. II, no. 7, pp. 27–36.

Schonberger, R. J. (1982) *Japanese Manufacturing Techniques: Nine Hidden Lessons in Simplicity*, Free Press, New York.

Schonberger, R. J. (1986) *World Class Manufacturing*, Free Press, New York.

Shingo, S. (1989) (English retranslation by Dillon, A.) *Toyota Production System from an Industrial Engineering Viewpoint*, Productivity Press, Cambridge, MA.

Turnbull, P. J. (1988) The limits to 'Japanisation' – Just-in-Time, labour relations and the UK automotive industry, *New Technology, Work and Employment*, Vol. 3, no. 1, pp. 7–20.
Voss, C. and Clutterbuck, D. (1989) *Just-in-Time: A Global Status Report*, IFS, Bedford.
Wickens, P. (1987) *The Road to Nissan*, Macmillan, Basingstoke.
Womack, J., Jones, D. and Roos, D. (1990) *The Machine that Changed the World*, Rawson Associates, New York.

10

CELLULAR MANUFACTURE AND THE ROLE OF TEAMS

David Buchanan

The aims of this chapter are to explore contemporary applications of autonomous work groups or self-regulating teams in manufacturing, and to assess the current and future potential of this approach. Recognizing that the concept carries different meanings in different settings, Mueller and Purcell (1992: 33) suggest that 'teamworking' has seven broad characteristics; their approach is based on a definition used by the General Motors/Opel car manufacturing group:

- The team works on a common task.
- The team has its own working space or 'territory'.
- Team members organize their own task allocations
- Members encourage and organize multiskilling.
- The team has discretion over work methods and time.
- The team has a leader or spokesman.
- Members can influence recruitment to their team.

This chapter has two main arguments. First, the autonomous work group approach has gained renewed popularity, and has found expression in more radical organizational configurations, as management see in employee 'empowerment' and 'responsible autonomy' the potential for increasing the flexibility and responsiveness required to deal with turbulent competitive and economic circumstances. Second, and despite this renewed popularity, the credibility of autonomous group working as a key manufacturing management approach is now under threat, from poor conceptualization, from a lack of systematic empirical evidence, and through international comparisons which appear to show the organizational superiority of Japanese car manufacturing methods.

It is important to recognize that, while Japanese companies use teamwork extensively in their manufacturing organization, the practice differs from the notion of autonomous work groups explored in this chapter. As MacDuffie explains,

> In Japan, teams in manufacturing plants . . . are fairly small, with six to eight members. In contrast with the well-publicized Scandinavian approach to assembly plant teams in the 1970s, jobs are still individually defined along a moving line. Team members do not carry out assembly tasks together, with rare exceptions. Similarly, Japanese teams do not function as autonomous administrative units responsible for a variety of personnel tasks.
>
> (MacDuffie, 1988: 14)

These teams are typically problem-solving quality circles, or 'continuous improvement groups', and do not directly affect the allocation of jobs on the shopfloor (Griffiths, 1992). However, 'teamwork' is falsely equated in some current management commentary with 'Japanese methods'. A further argument of this chapter concerns the need to resolve this confusion.

Autonomous groups and Group Technology are terms easily confused. The former is a *social* arrangement, as the above General Motors/Opel definition demonstrates. Group Technology is a *physical* arrangement which concerns the location of machinery on the shopfloor according to the product groups that can be manufactured on that equipment, and not according to machine function; in other words, all the widget-making machines are grouped in one area and all the sprocket-making machines are grouped in another, instead of putting all the turret lathes together, all the milling machines together, and so on. These contrasting approaches were developed independently in the mid- to late 1950s, in Britain and Russia respectively. A manufacturing facility that adopts Group Technology can achieve significant benefits in improved production scheduling and reduced materials handling, while still using traditional scientific management methods to design and allocate fragmented and specialized machine operating jobs; Group Technology does not require autonomous work groups. However, it is clear that the two approaches can be related, with self-managing teams of workers responsible for discrete product groups and their appropriate pools of equipment. Dawson (1991: 337) argues that Group Technology methods create a 'window of opportunity' for the improvement of Quality of Working Life through multiskilling and team independence. Contemporary developments in manufacturing methods involve just such a combined approach, also involving innovations in production control and inventory management techniques. The emerging trend thus concerns a *'socio-technical package deal'* of related and mutually reinforcing physical and organizational innovations. This vague concept will be explained and defined in more detail shortly.

FASHIONABLE AGAIN

The American management author and consultant, Tom Peters, raises the technique of autonomous group working to the level of organizational design principle, arguing that 'The modest-sized, task oriented, semi-autonomous, mainly self-managing team should be the basic organizational building block' (Peters, 1987: 296). Fresh management thinking? Twenty years previously, Miller and Rice (1967), in a work seldom consulted or cited today, advanced the case for effective organizational design based on considerations of task orientation and group affiliation. Rather than recklessly advocate a single approach, Miller and Rice also sought to identify settings in which closely knit, self-regulating groups could inhibit rather than facilitate organizational and technological change. They advocate what they call 'project type organization' – for manufacturing and administrative work – which involves the *temporary* formation of task-oriented groups which are dissolved when their tasks are complete. The social groupings to which employees belong, in a Miller and Rice project type organization, need not be consistent with task groupings as they would in a traditional autonomous work group approach. Other organizational arrangements are required if 'natural' social or occupational groupings do not meet affiliation needs. The power and subtlety of this analysis have been lost in much contemporary management commentary.

The concept of autonomous group working has its roots in the work of the Tavistock Institute in Britain in the 1950s (Miller and Rice were Tavistock consultants), and in widespread applications of the approach in Scandinavia in the 1960s and 1970s. Most reported applications seemed to imply significant benefits, in improved employee job satisfaction and in organizational performance. Group working was a key component in the tool kit of the influential Quality of Working Life (QWL) movement until the late 1970s when the approach seemed to lose popularity, and management attention shifted to other fashionable trends, to Japanese manufacturing methods, quality circles, Total Quality Management, and the effective exploitation of Information Technology. For example, Forslin (1992: 17) summarizes the shifts in management approach in the Volvo Engine Division in Sweden, reflecting the changing social and technical problems of the decades since 1950, moving from recruitment and labour turnover concerns to a focus on competitiveness, quality and responsiveness.

Why did the approach lose popularity? Autonomous groups and related job enrichment methods were aimed primarily at reducing labour turnover and absenteeism, and at improving job satisfaction, QWL and productivity. As unemployment worsened in the early 1980s, turnover and absenteeism ceased to be serious management problems, and QWL solutions may as a result have appeared less relevant. A second possible explanation is that the related organizational changes required to make self-regulating groups effective were seen as too radical by management conscious of how their own functions and prerogatives could be eroded. The labour problems of absenteeism and turnover hardly legitimated radical organizational redesign that also affected management roles (Buchanan, 1979, 1989). A third explanation is that the approach was not widely regarded as successful (possibly through the failure to introduce other supportive organizational changes). Valery (1974) estimated that over 1,000 job design experiments had been started in Sweden, but that most had failed, and many had been used only for publicity purposes. Which leads to a final consideration – was the approach genuinely popular? The number of accounts and analyses in the management literature may well have exceeded the number of applications on the ground, particularly as many commentaries recycled the same company experiences – Volvo, Saab, Atlas Copco, Philips, among others. The Swedish experience is still instructive in this respect, as evidence from the early 1990s suggests that the development of work organization and employee participation initiatives continues to be both patchy and uneven, both within Volvo and also across the Swedish economy as a whole (Cressey, 1992). The 'Swedish model' is not as common in that country as external accounts often tend to imply.

Autonomous group working has experienced something of a revival since the mid-1980s, if the management literature and Tom Peters are to be believed. Some larger corporations in America – Procter and Gamble, Digital Equipment, General Motors, Zilog, Hewlett Packard – appear to have sustained their interest in the approach, but avoided publicity on the grounds that their methods were considered to confer competitive advantage. Digital published internally an account of their experiment with autonomous teamworking at their Enfield plant in America (Perry, 1984). That experience was subsequently transferred (successfully and with local adaptation) to their small computer and semiconductor assembly business in Scotland (Buchanan, 1987; Buchanan and McCalman, 1989, 1992).

Digital describe their methods with the term, 'high-performance work system design'. Trendy new term for a well-established technique? Yes, in that the core of the approach remains the self-regulating or autonomous group able to make decisions and act on its own initiative within some more or less well-defined boundaries. No, in that the management motives are different and the related organizational changes are potentially more radical and also more difficult and time consuming to implement. Digital sought through high-performance work system design to enhance competitiveness, through improvements in product quality and reliability, and a reduction in time to market for new products. At a management seminar on high-performance systems design in Glasgow in 1988, a participant asked the speaker from Digital why he had revealed such a key aspect of the competitive advantage of the Digital assembly business in Scotland. 'Two reasons,' replied the Digital manager, 'First, because I know that you won't implement it, and second, because if you do it'll take you two or three years by which time we'll have moved on.'

The methods used by Digital, and by other organizations mentioned later in this chapter, suggest that we are no longer dealing with a straightforward management technique for improving job satisfaction. Many companies and commentators have come to see in autonomous group working a potentially powerful response to the competitive pressures, economic uncertainties, and technological changes of the 1990s. It is widely accepted (and thus rarely challenged) that organizations must become more quality conscious, more customer oriented, more adaptable and 'lean' and thereby more responsive, in order to compete and to survive. Autonomous group working appears on some evidence to confer such strategic benefits. In considering whether and how to implement an autonomous group working approach, the management motives now concern the search for improved competitive advantage, not for reduced absenteeism. This shift in motivation, it has been argued, now legitimates more radical approaches to work and organizational design (Buchanan, 1989). This shift in motivation, it appears, is why autonomous group methods have been enjoying renewed popularity since the mid-1980s.

IN SEARCH OF AN EMERGING PARADIGM

Fashionable or not, contemporary wisdom is remarkably consistent with respect to appropriate and effective responses to the competitive climate. If there is an emerging new paradigm (the 'socio-technical package deal' mentioned earlier) of manufacturing competitiveness, then it has at least four main, interdependent and mutually-reinforcing dimensions. These are the main dimensions of what has also been termed 'World Class Manufacturing' (Schonberger, 1986).

1. Adaptive structures

First, the competitive organization of the 1990s and beyond must have a structure that is flexible and adaptable. Once again, we can refer to previous research in support of this argument. Burns and Stalker (1961) argued that 'organic' management systems were more appropriate in dealing with environmental turbulence, and that 'mechanistic' or bureaucratic systems were too inflexible and more suited to stable environments. This argument has been restated more recently in more fashionable terms by Kanter (1983) who scorns what she calls 'segmentalist' (i.e. mechanistic) structures as too rigid, and advocates instead 'integrative' (i.e. organic) systems for dealing with a turbulent organizational environment. These new adaptive organizations are decentralized, delayered, and have also been described as 'adhocracies' (Toffler, 1970; Bennis, 1969). Reich (1983) argues that rapid changes in the technology of production and products demand the development of flexible systems of production to sustain competitive advantage. Increasing global market segmentation, informed and demanding consumers, complex and sophisticated technology, and increased turnover in tastes and fashions, Reich argues, mean that speed and flexibility of response are essential organizational characteristics. To the extent that autonomous group-based approaches to organizational design encourage flexibility, commitment, creativity and innovation, they potentially confer competitive advantage (Peters, 1987).

2. Self-managing teams

Second, the competitive organization must be based on socio-technical design principles with the self-managing team at the core. Socio-technical system design entails 'joint optimization' of the technical and social systems of the organization, and this in turn implies 'sub-optimization' of each system considered alone in the interests of the performance of the system as a whole. Much commentary on the application of Advanced Manufacturing Technology has suggested that it is not always appropriate to insist on investment in 'leading edge' or 'state-of-the-art' systems in order to be competitive (Ingersoll Engineers, 1987; Caulkin, 1990). Lawler (1986) identifies the features of 'new design plants', based on self-managing, multi-skilled teams, with harmonized payment and other working conditions, and a flat management hierarchy. Reinforcing the 'new paradigm' argument, Lawler claims that

> Overall, new design plants are clearly different from traditional plants in a number of important ways. Almost no aspect of the organization is left untouched. The reward system, the structure, the physical layout, the personnel management system, and the nature of jobs are all changed in significant ways. Because so many features are altered, in aggregate they amount to a new kind of organization.
>
> (Lawler, 1986: 178)

Developments in Advanced Manufacturing Technology have also led a number of commentators to advocate teamwork-based approaches to manufacturing organization design (Hirschhorn, 1984; Perrow, 1983; Hoerr, Pollock and Whiteside, 1986; Zuboff, 1988). This argument relies on the reasoning that, while advanced manufacturing systems require little manual intervention, the need for employee problem-solving and discretion is increased. Walton and Susman (1987) argue that the increased interdependencies in Advanced Manufacturing Technology systems, the speed and scope of errors, the pace of change, the sensitivity of performance to variation in employee attitudes and skilled intervention, and the greatly increased capital investment per employee all imply the need for flexibility, commitment, skill and motivation – to be achieved through multiskilling and teamwork. Cover stories in *Business Week* and *Fortune* magazines, carrying the headlines, 'Go team! The Payoff from Teamwork' and 'Who Needs a Boss?', respectively, give some indication of the extent to which autonomous group working has captured the attention of American management (Hoerr, 1989; Dumaine, 1990). Dumaine in fact describes the self-managed team as possibly '*the* productivity breakthrough of the 1990s' (1990: 40).

3. Considerate managers

Third, managers in the competitive organization must develop a different style, variously described as based on consensus rather than on direction (Pfeffer, 1992), or as entailing a shift from a culture of control to a culture of commitment (Walton, 1985). Walton and Susman (1987) speak of the need for lean, flat, flexible, innovative management and advocate strengthening the partnership between management and unions. They also argue for a reconsideration of the levels at which management decisions are taken. Lawler, adopting a different terminology, argues that

> High-involvement management is the competitive advantage available to countries with educated, achievement-oriented workforces who want to perform effectively, whose core values support participative decision making, and who can engage in substantial amounts of self-regulation. . . . The key question at this point is not whether high-involvement organizations will work. It is how to create such organizations.
> (Lawler, 1986: 215)

Mohrman, Ledford and Lawler (1986) similarly argue for the emergence of a new QWL/EI (employee involvement) paradigm. For a review of the literature supporting this argument, see Buchanan (1989).

4. Systemic changes

Finally, achievement of World Class Manufacturing status requires a combination of related technical, systemic and organizational changes (Schonberger, 1986). The search for more efficient manufacturing methods has begun to coalesce in what Drucker (1990) calls 'a new theory of manufacturing' which will characterize 'the post-modern factory of 1999'. Drucker's 'new theory' is founded on four concepts, concerning the development of modular or cellular organizational forms (i.e. Group Technology), statistical quality control, a strategic manufacturing management accounting, and a 'systems approach to the business of creating value'. A more detailed version of the post-modern factory can be found in the work of Parnaby (1988) of Lucas Industries. Parnaby has done much to publicize Just-in-Time and Manufacturing Systems Engineering (MSE) methodologies in Britain during the 1980s. His work has been widely disseminated in Production and Operations Management journals and is well known among production engineers. The Human Resource implications of the methodologies he describes have, however, only recently begun to attract interest (for example, Oliver, 1991), and have not been widely researched. Parnaby – with Drucker – argues for a 'total systems approach' to manufacturing organization design. This incorporates a range of related and supportive changes in many areas, including factory process control, organization structures, job structures, training programmes, process flow routeing, machine changeover procedures, communications systems, customer and supplier interfaces, project management for continuous improvement, capital expenditure authorization procedures, quality control, materials flow control, scheduling and planning systems, and inter-departmental relationships. Parnaby thus emphasizes that the development of Just-in-Time procedures is not merely 'the latest gimmick', is not a particular technique, and is not dependent simply on investment in new technology.

Drucker uses the term 'flotilla' to describe the multiple autonomous manufacturing cells which considered singly confer the benefits of standardization, and which combined offer flexibility for the organization as a whole. The core of the approach which Parnaby advocates also concerns the formation of a cellular organization structure, 'based upon natural people and machinery groupings around information or material flows' (Parnaby, 1988: 485); this again is a restatement of the Group Technology approach. This cellular structure should also be designed to facilitate the simplification of production control, the implementation of Just-in-Time workflow, the definition of cell and operator accountability for quality, the reduction in layers of organization hierarchy, flexible job structures and multiskilling, the elimination of traditional work orders and Work in Progress monitoring, team-based continuous improvement, and the introduction of relevant performance measures at manufacturing cell level.

This identification of a new paradigm for manufacturing organization and management is offered secure in the knowledge that it is partial and subject to other interpretations. The paradigm, in brief, advocates adaptive organization structures, plant lay-out based on Group Technology principles, with self-managing teams, supervised and co-ordinated by considerate managers who have also introduced systemic and supportive production engineering and control changes. Modern accounts are consistent in their advocacy of such an approach to manufacturing management, and this paradigm (if it can be so described) attracts little criticism in the management literature. This lack of critique is also striking; for where is the systematic empirical evidence to support the paradigm?

SHIFT IN MOTIVES, SHIFT IN METHODS: THE EVIDENCE

Research in this area is beset with methodological difficulties. How can changes in organizational performance be linked conclusively with specific organizational changes? The passionate claim that autonomous groups are more effective typically relies more heavily on intuition born of firsthand experience than on empirical evidence based on systematic research. As already noted, the implementation of teamwork is typically accompanied by inter-related and mutually-reinforcing changes in organization structure, management style, and production engineering and control systems. Such changes are invariably introduced during periods of technological and market changes which may themselves have furnished the motives for such developments. A number of accounts have indicated that performance metrics can be expected to deteriorate during the early stages of autonomous group implementation, as the organization enters what can in many instances be an awkward learning phase; performance subsequently improves as experience with the new systems and approach is gained. Leaving aside the question of which performance measures to use (these will presumably vary from site to site), we now have the added complications of determining the appropriate baselines and timeframes for an adequate evaluation study.

Recalling the motives behind these organizational innovations, managers cannot be expected to wait for conclusive research results before taking action to improve their competitive position. This is not surprising in this context, therefore, if such management action relies on limited evidence and much intuition. The available evidence is, however, certainly indicative, even if it is neither conclusive nor systematic. Accounts tend to be based on single company case studies rather than on cross-sectional analyses, and tend also to recount the 'success stories'. One suspects that the companies that gave access to researchers and permitted subsequent publication were those which had been successful with their organizational innovations in their own estimation.

Despite such obvious limitations, available accounts do have some positive attributes. They indicate the benefits and the problems – individual, managerial and organizational – of the approaches under discussion. They document experience from which other organizations can learn and offer a source of practical ideas for others to borrow and adapt. Published accounts appear also to indicate and confirm the trend towards the 'new paradigm' identified earlier. And these commentaries also highlight the gaps in theoretical knowledge and in effective practical application. So while the evidence we have may attract significant criticism, it should perhaps not be lightly dismissed. Assuming no new manufacturing management paradigm, team-based approaches to organizational design appear to have brought some organizations significant benefit; others can emulate this. Assuming on the other hand the reality of an emerging paradigm shift, these approaches may quickly become 'standard best practice' which organizations have to adopt merely to remain in business rather than to gain competitive edge.

What follows is a selective examination of contemporary accounts of autonomous group applications, providing a flavour of current practice and of the current available evidence. Volvo's car plant at Kalmar in Sweden was the first factory to be built around the concept of autonomous team-based assembly. The Kalmar plant has twenty assembly areas each with a team of fifteen to twenty people, with their own entrance, coffee area, showers and other facilities. Team members rotate jobs, are responsible for their own quality, and can vary their own work pace. Teams have access to all plant information through computer terminals in their work areas. Each team has its own 'buffer' area where work can be 'banked' to enable the group to extend its rest periods. The effectiveness of Kalmar has been challenged by rumours about the plant's inefficiency. Corlett and Sell (1987) argue that this is a myth, that Kalmar is Volvo's lowest-cost assembly plant and takes 15% fewer hours to make a car than the company average, and the number of man hours per car had been reduced 60% since 1977. In 1989 Volvo opened a new car plant at Uddevalla in west Sweden, extending the Kalmar approach with small teams building the complete car. In November 1989 Saab opened a new plant at Malmo in Sweden with a modular assembly system in place of the traditional conveyor belt (Lin, 1989). With absenteeism running at 25% in traditional plants, Saab decided to improve the working environment, offering more variety and responsibility in the work. Lin describes how, 'Decorated with greenery and garden ponds, the new factory is more reminiscent of a supermarket than a car plant.' Saab's 'car builders' work in teams of six to ten people, each with its own computer terminal to track stock levels. Each team member has twenty to sixty minutes to complete the production cycle, which could involve building a car door or fitting the transmission system. With a conventional assembly line, the work cycle was around two minutes.

Kroll (1989) describes the introduction of team-based methods at Smith, Kline and French Laboratories which manufacture 'blister packs' for pharmaceutical products. Due to the convenience and popularity of such packaging, extra manufacturing capacity has been required to meet a steady growth in demand since the mid-1980s. The trigger was, therefore, strategic.

The traditional packaging line required three groups of people: first, supervisors and direct labour, responsible for effective equipment operation, reporting to the blister packaging manager; second, engineers, responsible for changeovers, equipment settings and fault correction, reporting to the engineering group manager; and third, service operators, responsible for ensuring an adequate supply of materials and for cleaning lines between different runs, reporting to the packaging services manager. The blister packaging manager and packaging services manager reported to the packaging manager. Each of the three groups had its own separate responsibilities, but they were interdependent. When operators identified a minor modification to improve line efficiency, they had to get an engineer to implement it. Additional materials could not be acquired without a service operator. Breakdowns similarly could not be dealt with until an engineer was free. This combination of interdependence and separation of responsibility led to tension and frequent arguments. The managers each had their different objectives and priorities too. The organization structure inhibited efficiency gains, and the company started to look at other approaches.

As the company identified these problems in 1987, a new high-speed blister packaging line was being introduced. The decision was made to run this with a line team – a small group of highly trained operators with all the skills required to carry out the job. The team, with a leader and four members reporting to the packaging manager, became responsible for operating the equipment, carrying out in-process checks, performing equipment changeovers, and for correcting minor faults. The line team was responsible for everything except major breakdowns. The retraining involved a two-month programme which covered machinery, materials, documentation, and team development. The team leader was also put through a leadership skills course. The total cost of the training programme was £17,000. The resultant output was 2.6 times better than previously obtained from equivalent lines. Management felt this was due to a high level of motivation and to integrated team effort. The experiment on this one line was subsequently extended throughout the blister packaging department with similar performance improvements.

A similar, 'integrated working', approach has been developed at the Blue Circle Cement Works, at Dunbar in Scotland (Stevens, 1987). Blue Circle management wanted to develop a more flexible and skilled workforce, to remove 'them and us' barriers, and to stabilize income by reducing overtime. Integrated working had six characteristics. First, twenty-three teams formed the 'basic working unit'. Second, team leaders were given line responsibility; there were no deputies or chargehands, and team leaders have management training in leadership, communications and group skills. Third, 11% of the time of operators and craftsmen was devoted to training. Fourth, a monthly 'watchdog' works committee monitored the progress of the integrated working approach. Fifth, a 'stable income plan' replaced overtime, special payments, bonuses and hourly rates with an annual hours contract, with a 12.5% flexible hours component. Finally, the number of job grades was reduced to four, traditional demarcations were removed, and all personnel were expected to carry out any duty without direct supervision. Some problems remained. Employees complained that the flexible hours component of the annual hours contract was being used for routine work, and not for emergencies as originally intended. Redundancies (which cut the workforce from 480 to 250 in 1987) were felt to have been excessive. Complex clocking procedures were introduced. The speed of change in the company was generating anxiety.

However, there was considerable enthusiasm for the work variety, income stability, increased training, and improved management–employee relationships. The team leaders appreciated the freedom and the ability to use employee skills more effectively. More delegation meant that department heads had more time for planning. With respect to the goal of improved efficiency, the works manager is reported as saying that

> There is no doubt about that at all. I do not believe that we could have relied on the systems I knew ten years ago to implement the changes we made here. We needed a totally new approach to employee relations. There are still many unresolved difficulties, but our production per employee has risen from 1,900 tons per annum in 1983 to potentially over 4,000 tons today. Our productivity has increased dramatically, and this has not been due to new equipment and heavy investment alone.
>
> (Stevens, 1987: 12)

Peters (1987) describes the development of 'business teams' at the General Motors Cadillac Engine Plant in Livonia, Michigan. In this plant, everyone belongs to a business team of eight to fifteen people. The teams are responsible for their own work scheduling, training and problem-solving and they also develop their own quantitative performance indicators. Each team meets at least once a week as a group, and second-level supervision (general foremen grade) has been eliminated. First-level supervision (foremen) were reduced in number by 40%, and the role retitled 'team co-ordinator'. A pay-for-knowledge system encouraged everyone to learn virtually every job in the plant, and job specialization was virtually eliminated with the introduction of a single job category, 'quality operator'. The approach was implemented by a joint management–union planning team, involving hourly employees, working full time for a year on the plant organization and operation and on a plant operating philosophy.

Buchanan and Preston (1992) document the outcomes of the development of a manufacturing systems engineering approach in an engineering company – Moore Components. The approach combined Group Technology and the introduction of 'product autonomous cells' with a simplified shopfloor production scheduling and control system. It was intended that these changes would be implemented as a package, that the sixty-nine employees in the cell would operate autonomously, over three shifts, and that the role of the foremen would be transformed, 'from policeman to coach', in the words of one middle manager. The shopfloor lay-out and production control changes alone generated substantial benefits, including reduced manufacturing lead times, reduced costs and inventory, and increased capacity (without an increase in labour; the cell should have been running with ninety operators). However, cell autonomy was not developed, and the role of first-line supervision remained unaltered. The resultant dissatisfaction among cell members led to deliberate restrictions on further productivity improvements, and the company's aim of 'continuous improvement' was thereby threatened. In this instance, only the 'technical' elements in the manufacturing systems engineering package were implemented, while the 'socio-' components were overlooked. The immediate organizational benefits achieved were, nevertheless, significant.

The key factors blocking further immediate development of the approach in this instance appeared to lie with traditional management attitudes which condoned the maintenance of the traditional foreman's 'policeman' role. The intended change in job title to 'shift co-ordinator' did not take place, the training to help foremen adjust to their new role was minimal and ineffectual, and problems with the shopfloor production scheduling system encouraged foremen to sustain their traditional directive and interventionist methods. Employees on the shopfloor were frustrated by these developments, particularly as they had been allowed to experience freedom from supervisory interference for a period and as they had been told that this would be permanent. This case serves to illustrate the problems in implementing such a socio-technical package of changes, and Buchanan and Preston seek to demonstrate how the Human Resource policy environment of the organization can facilitate or inhibit such developments.

In each of the cases summarized here, it appears that some form of cell or group or team working generated some significant organizational benefits. However, these instances also reveal how a package of changes can generate a package of benefits and difficulties, rendering problematic the establishment of unambiguous casual links. Wall (1984: 28) argues that the productivity enhancement thought to flow from improved employee motivation typically relies instead on improved labour flexibility, mobility (around the production site) and ability to use initiative, and on reduced indirect costs. Kirosingh (1989) similarly argues that the costs and benefits of flexible working and teamwork have attracted little formal analysis, and that the main benefits derive from increased manufacturing capacity, reductions in the number of managerial and supervisory personnel, reduced inventory, and improved quality systems. The main costs of improving flexibility and teamwork arise from increased training and pay, and the input of management time. Despite the lack of firm evidence, however, Kirosingh estimates that 'the benefits of introducing flexible working practices outweigh the costs' (1989: 429).

OLD WINE, NEW WINE

What is new in all of this? Do these accounts truly represent a paradigm shift in manufacturing management methods, or merely a packaging shift, in which familiar ideas are presented with a fresh terminology?

The separate aspects of the approach are well documented, such as the emulation of Japanese manufacturing methods, the development in particular of Just-in-Time manufacturing methodologies, and attempts to introduce flexibility through increased use of out-of-craft and out-of-grade working on the shopfloor, and the adoption of multiskilled teamwork (Schonberger, 1982, 1986; Oliver and Wilkinson, 1988; Oliver, 1991; Buchanan and McCalman, 1989). The dimensions of manufacturing systems engineering are hardly novel. Cellular Manufacturing design is a well-established concept with roots in post-war Russian engineering where the concept of Group Technology was first applied (Mitrofanov, 1955). Group Technology, as explained earlier, is simply an approach to shop lay-out in which equipment is located according to product families which have similar processing requirements, and not according to machine function (Gallagher and Knight, 1973). Multiskilled teamworking, as already mentioned, has its roots in socio-technical system thinking as articulated in the 1950s and 1960s by consultants at the Tavistock Institute in London (Emery, 1963; Herbst, 1962; Rice, 1953, 1958; Trist and Bamforth, 1951; Trist *et al.*, 1963).

There is nothing new in the separate elements of the emerging paradigm, therefore. However, what do seem to distinguish contemporary applications are the *management motives* which, as argued earlier, appear to legitimate wider and potentially more radical approaches to organizational redesign, affecting management structures, functions and styles as well as shopfloor working practice. These single fads, single techniques, and single approaches are now being developed and applied in *novel configurations* in the search for improved performance and sustained competitive advantage. Peters (1989: 29) illustrates his claim that 'new organizational configurations are beginning to dot the landscape' with accounts of American companies that have dramatically reduced their management hierarchies while dramatically 'empowering' their shopfloor employees. In their analysis of company initiatives in the areas of performance-related pay, teamwork, training and employee involvement, Mueller and Purcell (1992: 29) argue that these cannot be justified or implemented singly, and that 'It is the integration of change initiatives with other aspects of organizational life which is the key to success. It is very rare for a single initiative, however well designed, to generate significant or lasting benefit.'

Three dimensions of these novel configurations are particularly significant in reinforcing and contributing to the potential effectiveness of autonomous group working. These concern rewards, training, and supervisory style.

It is a relatively straightforward matter to link material reward to attendance at work, timekeeping and physical output; these dimensions are easy to quantify. It is also relatively simple to specify job content as the basis for a job evaluation scheme where jobs are relatively static. However, traditional job evaluation becomes less relevant where job content is 'emergent', evolving over time as part of the individual's developing contribution to the organization and to the team (Mohrman *et al.*, 1986). How does a company measure and reward unquantifiable aspects of human performance such as initiative, commitment, adaptability, creativity? One answer to these problems lies in the development of skills-based pay, where individuals are rewarded for the business-relevant skills they have developed, independently of the particular jobs they perform at a given moment. Buchanan and McCalman (1989) describe the skills-based payment system designed, with shopfloor contribution, at Digital Equipment's plant at Ayr in Scotland. This approach was supported by 'behavioural assessment' in which team members appraised each other's contribution to the work of their groups. Mohrman *et al.* (1986: 108) offer evidence and argument supporting the shift from standardized, individual-based, secret, job-based pay to skills-based, egalitarian, open and flexible payment systems. Citing a number of companies which have introduced skills-based payment systems, or 'applied knowledge pay' (AKP) as it is known in Polaroid, *The Economist* (1991) points out that these developments are not motivated by the ideals of job enrichment and empowerment but by a desire to move away from traditional, rigid job classifications. Karlsson (1992: 22) outlines the development of a new payment system at ABB Distribution (a Swedish electrical equipment manufacturer) which combines traditional job evaluation with a skills-based element, and also recognizes individual and group performance.

Mohrman *et al.* (1986: 101) also summarize what they see as the emerging approach to training, based less on the individual job and focusing instead on the individual's understanding of and contribution to the work of the team and the organization as a whole. Employees expected to work in self-managing teams require teamworking capabilities, problem-solving skills and business understanding. Training must therefore go beyond basic job and technical skills.

Supervisors accustomed to traditional methods also require sophisticated training to turn them from policemen to coaches; recruitment, selection and promotion policies with respect to first-line supervision may have to be reviewed in this context. Contemporary reports of autonomous teamwork suggest that the role of supervision has in some companies been eliminated (Perry, 1984; Buchanan, 1987; Kirosingh, 1989), while in others it is being transformed from a traditional directive to a 'facilitative' and supportive role

(Wickens, 1987; Peters, 1987; Golzen, 1991). Peters (1987) argues for a shift in the first-level supervisor's span of control from ten to seventy-five, as the post changes from scheduling to coaching, from enforcing rules to facilitating, from planning to wandering, from providing solutions and ideas to helping teams to deal with their problems themselves. Adopting a typically uncompromising stance on these issues, Peters (1987: 300) states that 'if you do not drastically widen the span of control, and shift the supervisor's job content, the self-managing team concept will not work – period'.

This analysis, therefore, suggests a combination of some old wine with some new – packaged or systemic applications of known management techniques, with more wide-ranging organizational consequences, legitimated by a management preoccupation with finding competitive advantage through flexible and responsive teamworking. Several commentators in the 1970s saw job redesign techniques in future finding support in the social and technological changes which encouraged employee empowerment (see, for example, Davis, 1976); the pressures for change and the principles of good work design were seen as pointing organizations in the same direction. The anecdotal evidence now implies that management practice has caught up with this thinking; QWL seems to be consistent with the development of organizational capability to deal with technological innovation and competitive pressure. However, the reality of the 1990s may be neither as straightforward nor as optimistic as that.

THE CREDIBILITY FACTOR

This chapter has offered an optimistic account of the benefits of autonomous group working. The basic approach, supported by a range of other mutually reinforcing organizational changes, not only improves quality of working life, it is argued, but is an appropriate response to the turbulent technological, competitive and economic environment of the 1990s – and presumably beyond. There are, however, indicators that this optimism could be ill-founded, and that the credibility of the autonomous work group approach is under threat from at least three fronts.

First, despite around four decades of experience within the approach, the concept of autonomy is still poorly conceptualized and is in operation applied in a wide variety of different ways (Sauter, Hurrell and Cooper, 1989: xvi). In other words, when two people talk about autonomous groups, they may well not be discussing comparable phenomena, and their experiences may be wholly inconsistent and contradictory. The anecdotal evidence is therefore not cumulative and is handicapped in particular by the lack of comparability between cases. The concept of autonomy is always defined and operationalized locally, but there are no comprehensive, acceptable schema or frameworks that serve as a benchmark against which to assess individual applications. Gulowsen (1971) formulated a ten-dimension scale of group autonomy; Birchall and Wild (1974) proposed a similar framework, identifying six autonomy and four responsibility dimensions. These frameworks appear never to have been developed and applied, and are rarely cited (see Buchanan, 1979: 111–14). There is, therefore, no firm evidence to indicate how widespread the autonomous group approach might be; the assertion of popularity on which part of the argument of this chapter is based, relies on personal experience, a relatively small number of published accounts, and on a wholly subjective view that the approach is now receiving more management media exposure (in journals, books, training videos, and on television). These conceptual, theoretical and empirical gaps weaken the ability to advance compelling arguments.

Second, the basis of the potential benefits that are supposed to flow from autonomous group working are not well understood. The QWL movement stood at least in part on the argument that enriched jobs would improve performance, absenteeism and turnover through improving job satisfaction and motivation. But, as we have indicated, commentators more recently have suggested that flexible working has advantages that are independent of satisfaction. In the implementation of manufacturing systems engineering reported by Buchanan and Preston (1992), changes to shop lay-out, equipment moves and a new production scheduling and control system brought significant measurable benefits without the originally planned development of autonomous cells and facilitative supervision. The reported benefits in this and other similar cases, in the absence of a more sensitive causal analysis, may well arise from collateral changes to equipment and systems and not directly from a reorganization of work; this point has been recognized for some time (Wild, 1975: 20–1). It should also be noted in the context of this argument that these initiatives have invariably been championed by general and production line managers, and not by the Human Resource professionals whom one would expect to lead such developments (Storey, 1992).

Third, the credibility of the autonomous group approach is under threat now also from unfavourable comparisons with Japanese manufacturing methods. Put crudely, if Toyota and Nissan can make good-quality cars with Japanese methods more productively and more profitably than Volvo and Saab, what does that reveal about the socio-technical participative, flexible team-based work design methods developed by the Swedes? Japanese car plants just seem to roll out more cars with fewer defects at lower costs with fewer employees than their counterparts in America, Europe and Scandinavia. Japanese 'transplants' in America and Europe seem able to emulate that superiority (MacDuffie, 1988).

The international success of the Japanese motor car industry has been attributed to 'the Toyota revolution' of the 1950s and 1960s. This involved fresh approaches to materials handling, to the management of subcontractors, and the use of small-batch production with minimum inventory. It also involved the ruthless reduction of retooling and equipment adjustment times, the elimination of defects in the production process, and the use of minimum manning, multi-tasking, multi-machine operation, predefined work operations, repetitive short-cycle work, powerful first-line supervisors, and a conventional management hierarchy. This sounds like a classic scientific management approach. The ruthless and continuous search for improvement, known as *kaizen*, is one plausible explanation for Japanese manufacturing success. The focus of that continuous improvement, however, is typically a manufacturing facility designed on traditional scientific management terms, and radical organizational innovations are not involved.

Some media accounts now confusingly equate 'teamwork' with 'Japanese working practices'. Lorenz (1992) describes how Rover introduced a New Deal for its 35,000 employees involving various practices including multiskilling, flexibility, continuous skills improvement – and other 'Japanese manufacturing techniques like just-in-time delivery and team-working'. *The Economist* (1992) describes how Unipart reorganized production around machine cells run by small and flexible employee teams following a visit to a Honda component supplier in Japan. This confusion is in part a consequence of the lack of conceptual clarity surrounding the terminology of autonomous groups. Hoerr cites Haruo Shimada, an economics professor at Keio University in Tokyo, pointing out that in Japan,

> the team concept is not intended to increase workers' autonomy but to help them find out the problems in the production line so that no defective goods will be produced. In the US, workers tend to take participation as having a voice in all kinds of things that in Japan are determined by management and engineers.
>
> (Hoerr, 1989: 61)

Delbridge, Turnbull and Wilkinson (1992) argue that Japanese manufacturing organization and problem-solving methods increase the surveillance and monitoring of shopfloor workers, intensify work pressures, and reduce individual discretion with respect to working methods – in contrast with the prescriptions of the socio-technical school. They conclude that autonomy is superficial, that management dictate most production decisions, and that 'The result is far from a "humanized" workplace, where alienation is replaced with increased skill and input to the overall work process, but rather a highly regulated and regimented labour process with many of the characteristics of bureaucratic control' (Delbridge, Turnbull and Wilkinson, 1992: 102).

The Japanese concept of teamwork is thus in danger of being misunderstood. Japanese workers typically perform jobs that have been designed according to traditional scientific management principles. Japanese workers thus employed, form teams to examine how their otherwise traditionally run manufacturing facility can be continuously improved. This is quite different from the Anglo-Scandinavian concept of autonomous group working. Teamwork in Japan is therefore not the same as the autonomous group working discussed in this chapter.

In reality, the approaches which form the focus of this chapter were first formulated in Britain and have probably been most extensively applied in Scandinavia. It is clear from the description of 'teamworking' at the Nissan plant in Sunderland in Britain that their approach contrasts sharply with the autonomous group method. Nissan's personnel director, Peter Wickens (1987: 95), confuses the picture by concluding that 'teamworking and commitment are difficult to define but you know it when you see it; or perhaps more accurately, feel it. Teamworking is not dependent on people working in groups but upon everyone working towards the same aims and objectives.' Parker and Slaughter (1988) offer a different picture of 'teamwork' at the plant run by the Toyota–General Motors joint venture in California – New United Motors Manufacturing Incorporated (NUMMI; see also Huczynski and Buchanan, 1991: 302). This was publicized as highly successful and as an 'industry standard' approach for the automobile industry. Parker and Slaughter describe the approach developed at NUMMI as 'management by stress', with every motion and action timed meticulously to remove waste effort, reduce time and inventory, and streamline production continuously. The first team members were engineers and supervisors chosen by management and their first job was to break down every task in a conventional 'time and motion' analysis leading to a detailed written specification of how each team member should perform each job. This scientific management precision again contrasts sharply with socio-technical systems design methods.

Hammarstrom and Lansbury (1991) draw the obvious comparisons between 'Toyotism', Taylorism and Fordism. Why might such Japanese methods be effective in Britain? They point out that the workplace pressures are offset by high pay and job security, and that Japanese operations are typically found in areas of high unemployment and low or contained unionization. Also, in contrast with this Japanese success, the Swedish motor manufacturers have been in difficulty. In 1990 Saab sold its car business to General Motors (which in turn saw its market collapse in 1991–2). Volvo too saw sales tumble in the late 1980s and early 1990s and at the time of writing was exploring joint venture possibilities with other European car manufacturers (such as Renault), and was even (late 1992) planning to close the new autonomous teamwork-based plant at Uddevalla. Hammarstrom and Lansbury also point out that the Volvo approach is restricted to Sweden; Volvo in Belgium operates traditional assembly plants in areas with high unemployment and where union pressure for alternative methods is less than typically found in Sweden.

The apparent effectiveness of the manufacturing methods of Toyota and other Japanese car companies carries profound symbolic import. Fashions in management are easily followed – and equally easily discounted. For manufacturing managers looking for reasons to discount autonomous group methods, the Toyota–Volvo comparisons may provide them. In the absence of good evidence, manufacturing managers can be expected to take the Toyota experience as indicative of the weakness of the Scandinavian socio-technical approach. Hammarstrom and Lansbury (1991: 87) thus conclude that 'For managers in many countries, the Toyota approach appears to be a more "natural" and "safe" method of production. It seems likely that only management faced with shortage of labour, union pressure and reduced wage differentials will be forced to be as imaginative as Volvo has been at Uddevalla.' They raise the painful question, 'will the Volvo model for car production and work organization prevail?' (1991: 87). The motor car industry and its market position are of course unique, and it is inappropriate to generalize for that one industry experience across manufacturing as a whole. Will manufacturing managers in other sectors accept this argument and continue to develop their own team-based organizational approaches, or will they opt for the natural and the safe?

The key questions for the future, therefore, concern how the thinly substantiated credibility of the autonomous group approach and its sociotechnical systems underpinning can survive the continuing lack of conceptual clarity, the demonstrable lack of rigorous and strong supporting empirical evidence, and the invidious comparisons with Japanese manufacturing methods. These questions imply an interesting research agenda which will require organizational as well as academic commitment if they are to be addressed effectively.

One major addition to the research agenda must concern the role of cultural setting in influencing the acceptability and effectiveness of approaches to work organization. The concept of 'World Class Manufacturing' may imply the emergence (or even convergence) of 'international best practice'. This, however, would ignore the effects of the cultural, political and historical context in which management styles and organizational practices have developed and are applied (Ramsay, 1992). Political and historical developments in Sweden and Japan offer stark contrasts. Management motives and styles are different; workforce expectations and aspirations are different; the terminology of 'teamwork' denotes quite differently structured and differently driven practices across the two cultures. The transfer of management practice from one culture to another carries well-rehearsed dangers, even within the same sector. Contemporary cross-cultural productivity comparisons tend to ignore these factors, or treat them as interesting but relatively unimportant.

It has been suggested here that socio-technical approaches may be 'forced' on management by 'unfavourable' labour market conditions (Hammarstrom and Lansbury, 1991). However, there are signs that demographic changes are transforming Japan's traditionally dedicated and hard-working labour force, and that younger employees are showing interest in leisure and community pursuits clearly separated from working life, and in meaningful work which makes use of their capabilities. It will therefore be interesting to observe whether and how Japanese manufacturing managers change their approach to work organization in the mid-1990s, when faced with escalating employee recruitment, retention and motivation problems – precisely the factors which motivated Volvo to experiment with teamworking and to build Kalmar in the mid-1960s (Gyllenhammar, 1977; Forslin, 1992). Should Japanese managers be arranging study tours to Sweden in the second half of this decade in search of – for them – new and more effective methods?

REFERENCES

Bennis, W. G. (1969) *Organization Development: Its Nature, Origins and Prospects*, Addison-Wesley, Reading, MA.

Birchall, D. and Wild, R. (1974) Autonomous work groups, *Journal of General Management*, Vol. 2, no. 1, pp. 36–43.

Buchanan, D. A. (1979) *The Development of Job Design Theories and Techniques*, Saxon House, Aldershot.

Buchanan, D. A. (1987) Job enrichment is dead: long live high performance work design!, *Personnel Management*, May, pp. 40–3.

Buchanan, D. A. (1989) High performance: new boundaries of acceptability in worker control, in S. L. Sauter, J. J. Hurrell and C. L. Cooper (eds.) *Job Control and Worker Health*, John Wiley, Chichester, pp. 255–73.

Buchanan, D. A. and McCalman, J. (1989) *High Performance Work Systems: The Digital Experience*, Routledge, London.

Buchanan, D. A. and McCalman, J. (1992) The VLSI story, in K. Legge and D. Gowler (eds.) *Case Studies in Organizational Behaviour and Human Resource Management*, Routledge, London.

Buchanan, D. A. and Preston, D. (1992) Life in the cell: supervision and teamwork in a 'manufacturing systems engineering' environment, *Human Resource Management Journal*, Vol. 2, no. 4, pp. 55–76.

Burns, T. and Stalker, G. (1961) *The Management of Innovation*, Tavistock Publications, London.

Caulkin, S. (1990) Britain's best factories, *Management Today*, November, pp. 60–89.

Corlett, N. and Sell, R. (1987) Organizational effectiveness and quality of working life: learning from abroad – Sweden. WRU Occasional Paper, no. 39, November, Work Research Unit, London.

Cressey, P. (1992) Worker participation: what can we learn from the Swedish experience? *P+European Participation Monitor*, no. 3, pp. 3–7.

Davis, L. E. (1976) Development in job design, in P. Warr (ed.) *Personal Goals and Work Design*, John Wiley, London.

Dawson, P. (1991) From machine-centered to human-centered manufacture, *International Journal of Human Factors in Manufacturing*, Vol. 1, no. 4, pp. 327–38.

Delbridge, R., Turnbull, P. and Wilkinson, B. (1992) Pushing back the frontiers: management control under JIT/TQM factory regimes, *New Technology, Work and Employment*, Vol. 7, no. 2, pp. 97–106.

Drucker, P. F. (1990) The emerging theory of manufacturing, *Harvard Business Review*, Vol. 68, no. 3, May–June, pp. 94–102.

Dumaine, B. (1990) Who needs a boss? *Fortune*, May 7, no. 10, pp. 40–7.

The Economist (1991) New ways to pay, 13 July, p. 83.

The Economist (1992) 'Unipartners', 11 April, p. 89.

Emery, F. E. (1963) Some hypotheses about the ways in which tasks may be more effectively put together to make jobs (Tavistock Institute of Human Relations). Reprinted in P. Hill (ed.) *Towards a New Philosophy of Management* (1971) Gower Press, Aldershot, pp. 208–10.

Forslin, J. (1992) Towards integration: the case of Volvo Engine Division, *P+European Participation Monitor*, no. 3, pp. 12–18.

Gallagher, C. and Knight, W. (1973) *Group Technology*, Butterworth, London.

Golzen, G. (1991) Let the shopfloor manage itself, *Sunday Times*, 6 January.
Griffiths, J. (1992) Honda reaches accord in UK, *Financial Times*, 9 November, p. 10.
Gulowsen, J. (1971) A measure of work group autonomy, in L. E. Davis and J. C. Taylor (eds.) *Design of Jobs*, Penguin Books, Harmondsworth, 1972, pp. 374–90. First published as *Selvstyrte Arbeidsgrupper*, Tanum Press, Stockholm.
Gyllenhammar, P. G. (1977) *People at Work*, Addison-Wesley, Reading, MA.
Hammarstrom, O. and Lansbury, R. D. (1991) The art of building a car: the Swedish experience re-examined, *New Technology, Work and Employment*, Vol. 6, no. 2, pp. 85–90.
Herbst, P. G. (1962) *Autonomous Group Functioning*, Tavistock Publications, London.
Hirschhorn, L. (1984) *Beyond Mechanization*, MIT Press, Cambridge, MA.
Hoerr, J. (1989) The payoff from teamwork, *Business Week*, 10 July, pp. 56–62.
Hoerr, J., Pollock, M. A. and Whiteside, D. E. (1986) Management discovers the human side of automation, *Business Week*, 29 September, pp. 60–5.
Huczynski, A. A. and Buchanan, D. A. (1991) *Organizational Behaviour: An Introductory Text*, Prentice Hall International, Hemel Hempstead.
Ingersoll Engineers (1987) *Technology in Manufacturing*, Ingersoll, London.
Kanter, R. M. (1983) *The Change Masters: Corporate Entrepreneurs at Work*, Unwin, London.
Karlsson, K. (1992) New industrial work methods at ABB Distribution, *P+European Participation Monitor*, no. 3, pp. 19–23.
Kirosingh, M. (1989) Changed work practices, *Employment Gazette*, August, pp. 422–9.
Kroll, A. R. (1989) New working practice for blister packaging lines, *Manufacturing Chemist*, November, pp. 33–8.
Lawler, E. E. (1986) *High Involvement Management: Participative Strategies for Improving Organizational Performance*, Jossey-Bass, San Francsico.
Lin, X. (1989) Saab gives its workers their heads, *Eurobusiness*, December, p. 17.
Lorenz, A. (1992) Rover drives for Japanese working practices, *The Sunday Times*, 5 April, p. 3.7.
MacDuffie, J. P. (1988) The Japanese auto transplants: challenges to conventional wisdom, *ILR Report*, Vol. xxvi, no. 1, Fall, pp. 12–18.
Miller, E. J. and Rice, A. K. (1967) *Systems of Organization: The Control of Task and Sentient Boundaries*, Tavistock Publications, London.
Mitrofanov, S. P. (1955) *Scientific Principles of Group Technology*, National Lending Library, Boston Spa (translated from the Russian in 1966).
Mueller, F. and Purcell, J. (1992) The drive for higher productivity, *Personnel Management*, May, pp. 28–33.
Mohrman, S. A., Ledford, G. E. and Lawler, E. E. (1986) Quality of worklife and employee involvement, in C. L. Cooper and I. Robertson (eds.) *International Review of Industrial and Organizational Psychology*, John Wiley, New York, pp. 189–216.
Oliver, N. (1991) The dynamics of just-in-time, *New Technology, Work and Employment*, Vol. 6, no. 1, pp. 19–27.
Oliver, N. and Wilkinson, B. (1988) *The Japanization of British Industry*, Basil Blackwell, Oxford.
Parker, M. and Slaughter, J., (1988) *Choosing Sides: Unions and the Team Concept*, Labour Notes, Detroit.

Parnaby, J. (1988) A systems approach to the implementation of JIT methodologies in Lucas Industries, *International Journal of Production Research*, Vol. 26, no. 3, pp. 483–92.
Perrow, C. (1983) The organizational context of human factors engineering, *Administrative Science Quarterly*, Vol. 28, pp. 521–41.
Perry, B. (1984) *Enfield: A High Performance System*, Digital Equipment Corporation, Educational Services Development and Publishing, Bedford, MA.
Peters, T. (1987) *Thriving on Chaos: Handbook for a Management Revolution*, Macmillan, London.
Peters, T. (1989) New products, new markets, new competition, new thinking, *The Economist*, 4 March, pp. 27–30.
Pfeffer, J. (1992) *Managing With Power: Politics and Influence in Organizations*, Harvard Business School Press, Boston.
Ramsay, H. (1992) Swedish and Japanese work methods: comparisons and contrasts, *P+European Participation Monitor*, no. 3, pp. 37–40.
Reich, R. B. (1983) *The Next American Frontier*, Times Books, New York.
Rice, A. K. (1953) Productivity and social organization in an Indian weaving shed, *Human Relations*, Vol. 6, no. 4, pp. 297–329.
Rice, A. K. (1958) *Productivity and Social Organization*, Tavistock Publications, London.
Santer, S. L., Hurrell, J. J. and Cooper, C. L. (1989) *Job Control and Worker Health*, John Wiley, Chichester.
Schonberger, R. J. (1982) *Japanese Manufacturing Techniques*, Free Press, New York.
Schonberger, R. J. (1986) *World Class Manufacturing: The Lessons of Simplicity Applied*, Free Press, Illinois.
Stevens, B. (1987) Integrated working at Blue Circle, *Industrial Participation*, no. 594, Summer, pp. 10–12.
Storey, J. (1992) *Developments in the Management of Human Resources*, Blackwell, Oxford.
Toffler, A. (1970) *Future Shock*, Pan Books, London.
Trist, E. L. and Bamforth, K. W. (1951) Some social and psychological consequences of the longwall method of coal-getting, *Human Relations*, Vol. 4, no. 1, pp. 3–38.
Trist, E. L., Higgin, G. W., Murray, H. and Pollock, A. B. (1963) *Organizational Choice: Capabilities of Groups at the Coal Face Under Changing Technologies*, Tavistock Publications, London.
Valery, N. (1974) Importing the lessons of Swedish workers, *New Scientist*, Vol. 62, 4 April, pp. 27–8.
Wall, T. (1984) What's new in job design, *Personnel Management*, April, pp. 27–9.
Walton, R. E. (1985) From control to commitment in the workplace, *Harvard Business Review*, March–April, pp. 77–84.
Walton, R. E. and Susman, G. E. (1987) People policies for the new machines, *Harvard Business Review*, March–April, no. 2, pp. 98–106.
Wickens, P. (1987) *The Road to Nissan: Flexibility, Quality Teamwork*, Macmillan Press, Basingstoke.
Wild, R. (1975) *Work Organization: A Study of Manual Work and Mass Production*, John Wiley, London.
Zuboff, S. (1988) *In the Age of the Smart Machine: The Future of Work and Power*, Heinemann Professional Publishing, Oxford.

11

WORKER RESPONSES TO NEW WAVE MANUFACTURING

Paul Adler

In most of the chapters in this book the primary emphasis has been upon the nature of managerial approaches in manufacturing. In this chapter the focus shifts and the main objective becomes to understand how workers have responded to the introduction of New Wave Manufacturing technology. How do the new wave technologies change workers' perceptions of their jobs? What are the effects on job satisfaction levels, skill and perceived job autonomy? These and other related questions are examined in this penultimate chapter largely through the analysis of workers' responses to the introduction of Flexible Manufacturing Systems (FMS).

Recent technological advances are revolutionizing small- and medium-batch manufacturing. Small-batch manufacturing plants with relatively low levels of automation are being catapulted to the leading edge of automation – despite a long-standing and widespread assumption that small-batch production precludes automation (see Woodward, 1958; Hayes and Wheelwright, 1979). FMS extend computer control from stand-alone Numerically Controlled (NC) machines to groups of four to twelve machines under centralized computer control. These systems incorporate ancillary tasks such as materials handling, tool management and (sometimes) inspection. FMS transform small-batch jobshops into quasi-continuous processes: once loaded on to a fixture and released into the FMS, a part may undergo dozens of machining and inspection operations without being touched by human hands and may reappear only for unloading an hour or more later. As of 1984, there were about 60 such systems in the USA, 100 in Japan, 25 in the Federal Republic of Germany, and 15 in Sweden (Edquist and Jacobsson, 1988). In the mid-1980s the capital costs per installation were typically about $10 to $15 million.

With their high levels of automation and task interdependence, FMS are particularly interesting contexts in which to study the theoretical issues of work organization and effectiveness. To date, only one in-depth study of workers on an FMS installation has been reported (Blumberg and Gerwin

(1984), also reported in Blumberg and Alber (1982) and Cummings and Blumberg (1987)); there are several other studies of FMS work organization, but they are not as systematic as the Blumberg and Gerwin (1984) study: Ebel (1985), Jaikumar (1986), Jones (1985), Jones and Scott (1986), Graham and Rosenthal (1985), Kohler and Schultz-Wild (1985), Schultz-Wild and Kohler (1985), Seppala, Tuominen and Koskinen (1985) and Toikka (1985). The study by Blumberg and his colleagues has received considerable attention since it revealed two disturbing results. First, this study found profound worker dissatisfaction, reflecting a severe lack of autonomy and skill variety. In terms of the Job Diagnostic Survey (Hackman and Oldham, 1980), the Motivating Potential Score of these jobs was only 60% of that of a normative sample of machine trades jobs. Second, this study found that FMS often exhibit rather poor system performance, with utilization levels of only about 50–60%. These results led Blumberg *et al.* to recommend that managers use semi-autonomous work groups as an antidote to the human and operational problems they found.

But are other FMS equally inhospitable to workers? Is teamwork the required antidote to the motivational and technical difficulties found in this case study? In order to explore these issues, this chapter presents case studies of two other FMS and compares them to the case discussed by Blumberg and his colleagues. These new cases are particularly interesting because they have very similar technological profiles – indeed, they were designed and supplied by the same vendor in the same timeframe – but they have very different work designs: one retained the traditional division of labour that characterized the installation studied by Blumberg *et al.* while the other adopted the teamwork philosophy proposed by these same authors. Interview, observation and questionnaire data on these cases suggest some possible deficiencies in the prevailing theoretical frameworks and some fruitful lines of future research.

BACKGROUND: ISSUES IN RESEARCH OF AUTOMATION AND WORK

Automation and work requirements

The historical relationship between FMS and NC cannot help but shape the FMS research agenda. There is an abundant literature on the effects of the transition from conventional to NC machine tools on work content and organization (reviewed by Adler and Borys, 1989). Much of this literature highlights the 'de-skilling' and 'degradation' trends discussed by Braverman (1974). Braverman argued a dual thesis: (a) whatever the potential for more challenging jobs that may be intrinsic to new technologies, the struggle between workers and managers for workplace and work-pace control leads managers to adopt implementation modes that deprive workers of their autonomy; and (b) that competitive pressures and the profit motive lead managers to attempt to reduce costs and thus encourage implementation approaches that reduce both worker skill requirements and wage levels. This perspective would lead us to expect that workers would experience a decline in the quality of work as they went from conventional to NC machines and from NC machines to FMS.

Against Braverman's emphasis on social conflict and contrary to his prognosis of progressive de-skilling and degradation, there are three other well-articulated positions. The 'upgrading' position associated with industrialization theories (Kerr *et al.*, 1964) and theories of the post-industrial society (Bell, 1973) offers a prognosis based on the superior productivity of automation when associated with skilled users. This perspective would lead us to expect work requirements to be continually increasing with the passage from conventional to NC to FMS.

Another, related, theoretical tradition argues that there may be de-skilling effects in the early phases of mechanization, but that automation as a distinct, more advanced, phase holds the promise of job upgrading (Blauner, 1964; Woodward, 1958). From this perspective, we would expect skill requirements and the quality of jobs to follow a curvilinear path as machining progresses from conventional machine tools' stand-alone use in batch production to the more assembly-line-like role of NC operators, and from there to the continuous process of the FMS operating under human supervision.

Finally, against all these 'deterministic' theories, many theorists argue that the impact of automation on work requirements reflects many other contingencies such as management strategies, the state of product and factor markets, the power of contending actors, the social construction of skill categories (for example, Kelly, 1985; Child, 1987). This perspective focuses on the difficulty of making any compelling generalization about automation's impact on work requirements.

The study by Blumberg and his colleagues seemed to suggest that FMS were indeed fulfilling Braverman's prognosis. However, they report data for only one installation – I shall call it Blum Corp. Are other FMS equally inhospitable? To explore this question, I asked workers on two other FMS – I shall call them Team Corp. and Neotrad Corp. – to evaluate their current job demands and to compare them with those experienced in their previous jobs. The comparison with previous jobs was particularly interesting since several workers had worked on conventional machine tools or NC machine tools prior to their current jobs.

Job characteristics and work outcomes

A second set of concerns raised by FMS centres on the relationship between job characteristics and work outcomes, in particular the role of autonomy and teamwork in work satisfaction and work group effectiveness. The association of autonomy and satisfaction is deeply ingrained in our thinking. Both 'labour process' and 'job characteristics' models of work accord autonomy a central place.

First, Braverman's labour process analysis of the de-skilling of the machinist is based primarily on the idea that conventional machinist's jobs are de-skilled and/or degraded when, with the shift to NC, operators must share machine control responsibility with part programmers. This has led more recent research (Kelley, 1987) to attempt to test empirically the de-skilling hypothesis by ascertaining whether NC operators get to do any of their own programming or editing. (Kelley's data suggest that they usually do at least some editing.)

Second, the job characteristics model advanced by Hackman and Oldham (1980) makes autonomy a key factor in work satisfaction and effectiveness. Their model postulates (a) a causal chain running from 'core job characteristics' to 'critical psychological states' and from these psychological states to subjective and objective 'outcomes', and (b) some 'moderator' variables that may mediate these two causal links. The job characteristics model highlights the importance of autonomy through the assessment of job characteristics that are all defined in terms of the individual worker. The Motivating Potential Score of jobs – calculated as ((skill variety + task identity + task significance) / 3) × autonomy × feedback – is the variable that underlies both critical psychological states and outcomes. Not only autonomy, but also task identity, task significance and feedback are all assessed by items that are exclusively individual in scope. If, instead of performing a complete production cycle, the worker is specialized and performs only a subset of that cycle, and if through this specialization the worker is tied into a network of interdependence with other workers, then almost all the job characteristics when measured at the individual level could be expected to suffer.

But what if this individual autonomy were not central to work satisfaction? Proponents of semi-autonomous team job design acknowledge that individual autonomy may not be the *sine qua non* of work satisfaction. But they project that autonomy requirement on to the group level. Thus Hackman and Oldham (1980: 171–2) propose that the group tasks will be intrinsically motivating if there is a high level of the *group*'s task variety, identity, significance, autonomy and feedback. Such group task characteristics will be optimal when there is a relatively high level of technical interdependence of individual tasks, technical uncertainty and environmental change – all of which characterize FMS operations – and indeed under such conditions self-regulating work groups are the theoretically optimal manner of organizing these tasks (Cummings and Blumberg, 1987). In this spirit, and as a result of the low levels of work satisfaction they found, Blumberg *et al.* recommend that companies implementing FMS experiment with alternative, team designs of work (see also Kohler and Schultz-Wild, 1985). In this recommendation, they follow a venerable tradition in socio-technical systems designs (Pasmore, Francis and Haldeman, 1982) as well as the recommendations of Hackman and Oldham (1980).

But will team design overcome the difficulties that their sample of workers experienced? If so, is teamwork the only way to overcome those difficulties? And what are the effects of teamwork on work group effectiveness? To explore these questions, I compared Blumberg's results with results from my own study of two sites with contrasting work organization philosophies: Team Corp. was one of the only teamwork FMS installations in the USA, and Neotrad Corp. had almost the same technology as Team Corp. but had a rather traditional work organization philosophy similar in general outlines to Blum Corp.'s although rather more enlightened in the details of its application.

With respect to the impact of job characteristics on work satisfaction and effectiveness, and as long as workers in these installations have reasonably high growth and social needs, Blumberg *et al.*'s discussion would lead us to a cluster of expectations regarding the relative rankings of the three installations. The job specialization in Neotrad Corp. should generate a lower Motivating Potential Score, worse psychological states and poorer outcomes than in Team Corp. On the other hand, since the degree of job specialization is a key job characteristic – being a major influence on all five score factors – Neotrad Corp. indicators of Motivating Potential Score, psychological states and key outcomes should all be at about the same level as Blum Corp.'s. Since there would appear to be a poor fit between the contingencies characteristic of FMS operations and the traditional job designs used at Blum Corp. and Neotrad Corp., these systems' scores should fall below the scores typical of workers on conventional stand-alone equipment. And finally, since Team Corp. implemented the recommended antidote to job fragmentation – team organization – its scores should be at or above these scores.

Notions of technology

FMS also pose conceptual and methodological issues relative to our concept of technology. On the conceptual plane, FMS 'force us to rethink our conventional notions concerning technological development' (Blumberg and Gerwin, 1984: 114). Despite Woodward (1958), they are designed for small-batch production, yet they are highly automated; and despite Hickson, Pugh and Pheysey's (1969) workflow integration scale, they are highly automated yet very flexible. As researchers, we need to understand this new form of automation so vital to the thrust toward greater product variety that many see as the hallmark of the emerging conditions of competition (Blackburn, Coombs and Green, 1985; Piore and Sabel, 1984).

Methodologically, however, Goodman (1986) argues that this level of characterization of technology and task is insufficiently fine-grained. He argues that we need more in-depth characterizations of technologies and tasks if we want to understand work and work group effectiveness. The two new cases are particularly interesting in that, while they were technologically almost identical, their tasks – as defined by schedule constraints and the type and number of parts being produced – were somewhat different: Team Corp.'s task was very similar to Blum Corp.'s but differed from Neotrad Corp.'s. My analysis will provide some evidence for the validity of Goodman's argument.

METHODS

Approach

As indicated earlier, my approach is based on comparative case studies of two FMS installations with divergent work organization philosophies with the installation studied by Blumberg *et al.* We can summarize these philosophies as follows:

- Blum Corp. had a work organization philosophy of the most traditional kind. Its basic objective was to minimize overall labour costs by job specialization. This work design corresponds to what Cummings and Blumberg (1987: 45) call the traditional work group.
- Team Corp.'s philosophy and innovative – maximize teamwork, workforce flexibility, worker satisfaction, motivation and learning. In principle, this design corresponds to Cummings and Blumberg's self-organizing group, although in practice the degree of autonomy was somewhat limited by schedule requirements.
- Neotrad Corp.'s philosophy could be described as neo-traditionalist – a policy we might describe as 'conservative innovation' allowed them to reap the cost benefits of specialized formal job assignments characteristic of traditional work groups yet nurture motivation through some informal flexibility in job assignments and through longer-term promotion opportunities.

The primary data come from a comparison of worker responses to a questionnaire (described below) distributed by Blumberg and his colleagues to workers at Blum Corp. with responses to the same questionnaire distributed to workers in the two other FMS. These data were augmented by interview and observational material. In the sections below, these data will be used to sketch comparative profiles of the three installations. The first section sets the contexts; the second examines the differences in work requirements between FMS and NC and conventional stand-alone equipment; the third discusses the FMS' operating conditions; and the final section compares the three installations in terms of the three main elements of the Hackman–Oldham job characteristics model. A discussion section will link the cases to the background literature reviewed in the previous section.

The aim of the comparison is not to test hypotheses – a sample of two would hardly be compelling evidence. But rather, I hope to add to our stock of descriptive material on these novel automation settings and to clarify and sharpen the questions we bring to this area of study.

Blumberg *et al.* data: Blum Corp.

The data presented by Blumberg and Gerwin (1984), Blumberg and Alber (1982) and Cummings and Blumberg (1987) were based on a questionnaire incorporating the Job Diagnostic Survey (JDS) (Hackman and Oldham, 1980) as well as items from other studies (Quinn and Shepard, 1974; Quinn and Staines, 1979; Blumberg, 1980; Emery, 1972; Rousseau, 1977; Walton, 1977).

They administered the questionnaire to eighteen of the twenty direct workers and supervisors working on two shifts on Blum Corp.'s FMS, which will be described in more detail in the next section. They compared these results with a normative sample of sixteen machine trades workers (Oldham, Hackman and Stepima, 1979) and, in other parts of the questionnaire, to a sample of 1,515 employed adults representative of all occupations in all industries in the USA analysed by Quinn and Staines (1979) and to a similar sample of 1,496 workers analysed in the Quality of Employment Survey (QES) by Quinn and Shepard (1974).

New data: Team Corp. and Neotrad Corp.

My data come from the same questionnaire as those of Blumberg and his colleagues (thanks to the kind co-operation of Gerwin) administered to the workers in the FMS in Team Corp. and Neotrad Corp. during 1986. I also had the opportunity to visit both sites and conduct interviews with several workers, engineers and managers.

The entire questionnaire was administered to all three shifts at both companies – a total of fifteen workers in Team Corp. and nineteen in Neotrad Corp. With management approval, each shift was approached as a group and a room was arranged where they could fill out the questionnaire during a shift break or after work. If those times were not convenient, we attempted to arrange other more convenient times. Within this context, participation was voluntary, and participants were assured confidentiality. The final response rate was fifteen out of fifteen in Team Corp. and nineteen out of twenty-one in Neotrad Corp. This questionnaire contained both closed and open-ended questions. The closed questions give us quantitative responses on both the stand-alone/FMS comparison and the augmented JDS assessment.

THREE CONTEXTS

Blum Corp. was a 'diversified American manufacturer with sales of $2 billion in 1980, which has a division producing tractors for which the major housings are machined on a flexible manufacturing system' (Blumberg and Gerwin, 1984: 116). The Blum Corp. FMS was purchased in 1972 at a cost of some $5 million. It was one of the very first FMS in the USA (Cummings and Blumberg, 1987: 41). As described by Jones and Scott (1986), two other researchers who studied this installation, this FMS project was seen by the company as a learning opportunity: 'Originally conceived as a joint venture between a machine tool manufacturer eager to develop expertise in these systems, but lacking floor space of its own, and [Blum Corp.], it was to help in the production of a new tractor model' (Jones and Scott, 1986: 5).

The Blum Corp. FMS produced six major housings for a new tractor line. Each housing occupied approximately one metre cube and weighed about one metric ton (Cummings and Blumberg, 1987: 41). The system consisted of ten machine tools and three load/unload stations over a floor area of approximately 100,000 square feet and linked by twelve tow chains for material handling carts.

The work organization on this system was of a traditional kind: the two-shift operation employed six load/unload operators at a low labour grade, six operators with NC machining backgrounds at higher grades, two tool setters, two mechanics, and two supervisors. The plant was unionized. There was no incentive pay for workers on the FMS since management believed that output was determined by technical factors rather than by worker effort (Cummings and Blumberg, 1987: 53).

The FMS installed by Team Corp. and Neotrad Corp. were very similar to each other in their technological dimensions, both having been built by the same vendor in the same timeframe to very similar specifications. They were both installed in 1985. Team Corp.'s system was built around four identical CNC machining centres, each with ninety-tool storage capacity, one Coordinate Measuring Machine (CMM, for automated inspection of parts' dimensions), two load/unload stations, two automatic work changers (or pallet parking areas) and three Automated Guided Vehicles (AGV, for part transport). Neotrad Corp.'s system was identical except for the addition of four extra CNC machining centres, one CMM and one AGV.

Beyond these similarities, however, Team Corp. differed from Neotrad Corp. in four main respects. First, the mix of motivations for the FMS investments differed somewhat. Interviews with managers at Team Corp. revealed that their FMS investment was encouraged by a corporate-level executive sponsor whose main concern was pushing the organization to learn about FMS technology so as to be able to use it elsewhere in the company. In the process of mobilizing support for the $15 million investment, two other motives became germane: to reduce costs and to project an image of technological dynamism to their customer, the Department of Defense. Neotrad Corp. was not a 'prime' contractor working directly for the DoD, but rather sub-contracted major segments of work from such prime contractors; it therefore competed much more directly on cost, and cost reduction was therefore its primary motivation. The Neotrad Corp. production manager also saw the FMS as a solution to the difficulty, commonly experienced in the metal-forming industry, of attracting and retaining skilled and motivated machinists who, in his words, 'were willing to go the extra mile' for effective operations.

A second difference was that Team Corp.'s FMS was located in a relatively new non-union plant (opened in 1983) with a policy of innovation in work organization – all its workers were salaried, for example. Neotrad Corp.'s installation was located in an old and unionized plant.

Third, the tasks in each of the FMS differed in significant details. Team Corp's FMS, like that of Blum Corp., produced a small number of large-size complex parts – in Team Corp.'s case, some twenty-five gear housings. In the Team Corp. case there had been persistent quality problems in the casting components delivered to the site. This created organizational stress because the FMS are less tolerant of variations in casting dimensions than conventional methods since they allow no on-the-spot adjustments to feeds and speeds. At Neotrad Corp. the FMS produced a large number of parts – over 500 – and these were smaller and simpler in both shape and machining requirements. On the other hand, Neotrad changed part designs more frequently, putting strain on the relationship between part programming and the off-line tape proofing operation. A foreman estimated that fewer than 1% of new part programs were correct the first time.

Finally, Team Corp.'s FMS had a particularly innovative work design. The workers on the FMS installation were organized as a team, with job rotation and vertical job enrichment. All the Team Corp. FMS positions, including control room operator, were rotated approximately weekly. The rotation schedule was not rigid – team members set it themselves – and employees were trained in the more difficult jobs before being rotated into them. Their compensation was based on a pay-for-knowledge scheme that gave salary increments for new skills acquired. Workers moved flexibly between jobs in crisis or overload situations. Not only were jobs rotated, but they were also considerably enlarged and enriched compared to either Blum Corp. or Neotrad Corp. The load/unload station operators, for example, had considerable discretion over when to respond to a system call to load or unload, and their jobs were enlarged to include deburring and quality inspection. Job enlargement was also reflected in the preventive maintenance responsibilities of the operators. In this job design, Team Corp. conformed very closely to the recommendations of Blumberg and his colleagues and to those formulated by the Manufacturing Studies Board (Manufacturing Studies Board, 1986; see summary in Walton and Susman, 1987).

The Team Corp. FMS was located in a greenfield plant. To be hired into the plant, candidates were required to take a twelve-week training course at a local technical college in their own time and without the guarantee of a job. The plant had only three pay classifications for direct labour, and the FMS workers were all paid at the highest of the three. Given the amount of cross-training the Team Corp. workers received, there was some debate within management as to whether they should be paid at a yet higher level. But management decided against that policy because, with a higher grade, the plant's job bidding system would have restricted the recruitment of FMS workers to workers with the most seniority, whereas management wanted to recruit on the basis of criteria such as learning speed, disciplinary record, motivation, and peer respect. (As FMS workers acquired control room operator skills, however, it was anticipated that they would probably be promoted to a plant technician grade.) To further emphasize the motivational challenge of the FMS, workers were on rotating shifts for the first year and a half, so that to join the FMS staff, workers had to give up their seniority-based shift privileges. This had the added advantage of giving all the FMS staff experience on the first shift when most of the debugging was done. On each shift team there were three experienced NC operators, an experienced equipment maintenance person, and usually one person with tooling experience.

Unlike Blum Corp. or Team Corp., the original group of Neotrad Corp. FMS workers were new to the plant; since other departments in the plant were also hiring at the time, the union did not object. Neotrad Corp.'s work organization was very similar in general outline to Blum Corp.'s: work roles were specialized into operators, control room personnel, loaders, mechanics and tool setters, and the job descriptions of the more skilled jobs were written so as to ensure that key people would not be 'bumped' by workers from elsewhere in the plant who had more seniority but no experience on the FMS. In the opinion of the Neotrad Corp. production foreman, some of the positions deserved to be paid at a rate above that of the Class 'A' machinists, but the personnel department balked at the thought of creating a new grade to which machinists elsewhere in the plant could aspire. The load/unload station workers were initially classed significantly lower than the lowest machine shop grade, because management assumed that their jobs were basically labourer-type jobs whose responsibility extended no further than bolting and unbolting parts to fixtures. But as the FMS operations ramped up, management revised their assessment in light of the responsibility required of these workers for very precise part positioning, for quality control (the load/unload operators were the first to see parts as they came off the system and were therefore well placed to immediately notify the control room of any discrepancies) and for timely performance. As a result, their classification was brought up to the lowest machine shop level.

Not only did Neotrad Corp. display some flexibility in their implementation of the FMS, but within the rather traditional job specialization system that they maintained, workers were given significant longer-term training and promotion opportunities. By informal policy, priority in filling open positions within the FMS was given to promotions from within the department. Workers could and did progress from off-line deburring to load/unload, to operator, and even to control room positions. The control room positions were filled by former machinists, because, according to the production manager, computer systems' personnel tended to get absorbed by the software issues and lose sight of the metal-cutting operation that was their *raison d'être*.

COMPARING FMS AND STAND-ALONE SYSTEMS' WORK REQUIREMENTS

I identified five workers in Team Corp. whose prior jobs were on conventional or NC machine tools and a total of four in Neotrad Corp. whose prior jobs were on NC machine tools. (Other workers came from a variety of backgrounds, almost all in metal-working or mechanical occupations, and on average the workers in the two installations had similar education levels and age structures.) When comparisons were made between the workers' experience of FMS and their previous jobs on conventional and Numerical

	Team Corp. (n = 15) Mean (SD)	Neotrad Corp. (n = 19) Mean (SD)
Skill	4.33 (0.58)	3.79 (0.50)
Experience	4.13 (0.91)	3.90 (1.29)
Sense of responsibility	4.40 (0.91)	4.00 (1.20)
Teamworking	4.87 (0.35)	4.11 (1.10)
Training	4.23 (1.03)	3.84 (1.43)
Concentration	4.00 (1.25)	3.95 (1.31)
Interaction with support functions	4.27 (1.10)	4.00 (1.14)

Figure 11.1 Comparing FMS with work in stand-alone conditions
The question asked was 'How would you compare working on the FMS and on your previous job?' Working on FMS requires: 1 = a lot less, 2 = a bit less, 3 = about the same, 4 = a bit more, 5 = a lot more skill, experience, etc.

Control equipment, three results emerged (see Figure 11.1). First, on all dimensions, including skill levels, sense of responsibility, teamwork, required concentration and interaction with support functions, the FMS conditions produced higher scores. That is, working on FMS was perceived as requiring *more* of these attributes. Second, the increase was greater in Team Corp. than in Neotrad Corp. in every dimension except concentration and ability to handle boredom. Third, the transition to FMS from conventional equipment seemed to be about as demanding in most dimensions as the transition to FMS from NC: on most dimensions, the results are identical; the increase in 'experience' demands seems greater for former NC operators; the increase in 'interaction with support functions' seems greater for conventional machinists.

These survey data suggest a rather positive interpretation of the shift to FMS. This interpretation was supported by several interviews in both companies, in which none of the workers expressed any desire to return to work on stand-alone equipment.

Job characteristics seven-point scale (1 = low, 7 = high)	Team Corp. (n = 15) Mean (SD)	Neotrad Corp. (n = 19) Mean (SD)	Blum Corp.* (n = 16) Mean (SD)	Normative sample** Mean (SD)
Skill variety	5.44 (1.25)	4.88 (1.52)	3.65 (1.81)	5.08 (1.21)
Task identity	3.96 (1.89)	4.33 (1.70)	4.23 (1.95)	4.92 (1.30)
Task significance	6.04 (0.97)	5.74 (1.01)	5.66 (0.84)	5.61 (1.19)
Autonomy	4.93 (0.94)	5.09 (1.40)	4.04 (0.94)	4.93 (1.34)
Feedback from job	5.44 (1.00)	5.58 (1.07)	4.47 (1.68)	4.92 (1.15)
Motivating Potential Score	142.44	153.92	80.30	135.81

Figure 11.2 Results using the job diagnostic survey
* Data from Blum Corp. are adapted from Blumberg and Alber (1982) and Blumberg and Gerwin (1984). Supervisors have been excluded so as to allow a more direct comparison. ** The normative sample quoted is the JDS sample of machine trade workers (n = 16) (Oldham, Hackman and Stepima, 1979)

COMPARING JOB CHARACTERISTICS MODEL VARIABLES

In this section, Hackman and Oldham's (1980) Job Diagnostic Survey is used as a 'probe' to identify key facets of workers' experience. The results are reported in Figure 11.2. Interpreting these results, however, will require some digressions on the nature and limits of the probe instrument itself.

Job characteristics

Of the job characteristics variables, the distribution of scores for skill variety and task significance conforms approximately to the expectations of the background literature, while the three other variables and the total Motivating Potential Score do not.

Skill variety
The skill variety results for Team Corp. reflect the job-rotation policy in that installation. If Neotrad Corp. workers experienced a higher level of skill variety than Blum Corp. workers it was because, despite their equally narrow formal job descriptions, the informal work organization allowed Neotrad workers the use of a broader variety of skills through voluntary job switching.

Task identity
This was relatively low for all three installations. Two factors explain this result. First, the JDS questions measuring task identity focus exclusively on the identity of individual tasks. Compared to the 'pooled' interdependence characteristic of the stand-alone machine tools that are the predominant technology used by the normative sample, work on an FMS has a 'reciprocal' interdependent character within the work group (see Thompson (1967) on pooled, sequential and reciprocal interdependence). Second, as we have already seen, this interdependence extends to other shifts, support groups and suppliers.

Task significance
Task significance – the degree to which the job has a perceptible impact on the lives or work of other people – is somewhat higher in Team Corp. than in Neotrad Corp., Blum Corp. or the normative sample. Perhaps some of this difference reflects a Hawthorne effect, since Team Corp.'s use of an innovative work organization in combination with a novel technology had put it on the itinerary of more researchers than Neotrad Corp. or Blum Corp. Interviews also revealed that Team Corp.'s senior management was constantly tracking the delivery of the FMS parts, since these parts were in short supply, whereas Neotrad Corp.'s FMS shipped to a sizeable finished goods inventory.

Autonomy
Given the expectations derived from the background literature, the results for autonomy are surprising on three counts. First, it is surprising that Team

Corp. workers did not experience a higher degree of autonomy than Neotrad Corp. workers. As we have already seen, however, work on the FMS is closely interdependent within the FMS team, with other shifts and functions within the plant, and with suppliers. Team Corp. workers' sense of autonomy was perhaps also impaired relative to Neotrad Corp.'s by the fact that, as already pointed out, Team Corp.'s FMS was chasing its schedule. Second, if this interdependence is so constraining, it is surprising that Team Corp. and Neotrad Corp. did not fall below the normative sample composed primarily of workers on stand-alone machines. This result serves to remind us that even machinists on stand-alone equipment are typically tied into a complex web of interdependencies linking them to setters, the tool room, schedulers, and other support functions. This result reinforces that of Figure 11.1: even when their machines are not driven by NC (and only a small minority of the US stock of machine tools are NC), the autonomy of the journeyman machinist as depicted by Braverman is largely mythical. The third surprise in this data is that Neotrad Corp. and Blum Corp. workers experienced quite different levels of autonomy despite the similarity in their degree of job specialization. My hypothesis is that the experience of autonomy was shaped by the fine-grain texture of work experience and worker/supervisor relations. The voluntary versus involuntary character of job switching is one difference in texture that could give rise to different levels of perceived autonomy.

Feedback from job

While the Blum Corp. workers gave this variable a lower score than the normative sample, both Team Corp. and Neotrad Corp. workers scored it higher. This seems to reinforce the point made immediately above: this variable depends a great deal on supervisory style. It may also be the result of Blum Corp. workers' dissatisfaction with resource adequacy: without adequate resources, one's performance on the job does not reflect one's own efforts.

Motivating Potential Score

The net effect of the preceding five variables is that Team Corp. and Neotrad Corp., despite their very different job designs, offer similar degrees of motivating potential – a potential comparable to that of the normative sample – whereas the Blum Corp. installation, despite its approximate similarity with the technology of both Team Corp. and Neotrad Corp., and despite its similarity with Neotrad Corp.'s job specialization, offers much less motivating potential than the normative sample jobs.

The similarity of Team Corp. and Neotrad Corp. Motivating Potential Scores raises, however, a methodological question about the JDS: the JDS questions underlying the scores are exclusively at the individual level. The autonomy of the FMS team is not given any weight in the calculation of the Motivating Potential Score. We might therefore wonder whether Team Corp. workers taken as a team experienced higher job characteristic scores, and whether, as a result, the JDS estimate of Motivating Potential Scores was biased. Blumberg *et al.* included an extra question in this part of the questionnaire, asking respondents about the team's autonomy in deciding how to do its work. Contrary to expectations, Team Corp. scored a little lower than Neotrad Corp.: 4.66 (SD = 1.35) for Team Corp. as compared to 4.83 (SD = 1.65) for Neotrad Corp. The explanation for this result should now be clear: while Team Corp. workers have greater *initiated* interdependence than Neotrad Corp., they also experience greater *received* interdependence in the form of problems with casting suppliers and schedule pressures (Kiggundu, 1981). Autonomy correlates positively with the former but negatively with the latter. By contrast, the chief form of received interdependence at Neotrad Corp. was in proofing new parts, which was conducted off-line on separate machines, so difficulties here did not impinge as much on the experienced autonomy.

Hackman and Oldham's model identifies three 'critical psychological states': experienced meaningfulness of work, experienced responsibility and knowledge of results. On all three, Neotrad was above and Blum was below the normative sample level. Team Corp.'s team organization did not, however, enhance its members' psychological states relative to Neotrad. These results confirm the logic of the JDS: the psychological states parallel the Motivating Potential Score results shown in Figure 11.2.

Key outcomes

The primary outcome variables measured by the JDS reflect the general configuration of psychological states (see Figure 11.3).

Internal work motivation scores for Team Corp. and Neotrad Corp. were similar and above the normative sample. This buttresses the interpretation

Outcomes	Team Corp. Mean (SD)	Neotrad Corp. Mean (SD)	Blum Corp. Mean (SD)	Normative sample Mean (SD)
1. Internal work motivation (seven-point scale)	5.79 (0.82)	5.81 (0.58)	3.78 (1.07)	5.59 (0.80)
2. Satisfaction with challenge (four-point scale)	3.04 (0.47)	3.11 (0.54)	2.48 (0.70)	3.00 (0.68)
3. Satisfaction with promotion (four-point scale)	2.89 (0.60)	2.98 (0.81)	1.67 (0.80)	2.46 (0.86)
4. General satisfaction	5.09 (1.33)	5.40 (0.77)	4.50 (1.27)	4.91 (1.08)
5. Satisfaction with comfort (four-point scale)	3.06 (0.23)	3.11 (0.33)	2.55 (0.49)	2.87 (0.57)
6. Utilization of valued skills (five-point scale)	3.40 (2.03)	2.05 (1.81)	1.38 (1.03)	3.27 (1.29)
7. System utilization (%)	50	90	50–60	15–25

Figure 11.3 Key outcomes

advanced earlier that as far as the workers are concerned, the team organization of Team Corp. was not a decisive feature in distinguishing it from Neotrad Corp. Blum Corp.'s weak showing is congruent with its image as it has emerged in our discussion so far.

The Blumberg *et al.* questionnaire included several questions that assessed workers' satisfaction with two key dimensions of growth: *challenge* and *promotion possibilities*. On both these dimensions, we find a pattern similar to internal work motivation. These results are somewhat surprising: one might have expected these scores to be greater in the team organization with job rotation and cross-training. The puzzle is perhaps resolved by examining the temporal structure of the FMS teams' skill formation process. At Team Corp. the challenge of mastering all the jobs except control room operations had already been met, and extensive training for the system control positions was a near-term objective only for one or two Team Corp. workers. By contrast, job specialization at Neotrad Corp. left a broader range of training needs unmet. As a result, scores for both perceived challenge and perceived promotion possibilities were as high in Neotrad Corp. as in Team Corp.

General satisfaction reveals the same pattern as the preceding variables, but the gap between Neotrad Corp. and Team Corp. widened here. Was this widening the result of the disappointment of higher expectations built up in Team Corp.? Another hypothesis is that the exceptionally high system utilization level in Neotrad Corp. (discussed further below) provides a high level of satisfaction independently of the critical psychological states hypothesized to underlie general satisfaction.

The Blumberg *et al.* questionnaire also captures two other outcome variables. The ranking for *comfort* reflects the rankings seen above, while the ranking of *skills utilization* reflects much more directly the narrower division of labour in Neotrad Corp. and Blum Corp. that left many workers feeling under-utilized.

What can we say about the more objective outcome, *system utilization*? On this score, the ranking clearly separates three levels – first Neotrad Corp., second Team Corp. and Blum Corp., and finally the normative levels characteristic of stand-alone machines.

The much lower levels of system utilization of stand-alone machines are primarily the direct effect of technology: their utilization ratios (calculated as the ratio of cutting time to available time) are notoriously low because so much time is spent in set-up, materials handling, positioning, tool changes and downtime. These factors are typically compounded by the poor organization and scheduling practices characteristic of many machine shops.

The disparity between Neotrad Corp. on the one hand and Team Corp. and Blum Corp. on the other is due not to technology so much as task requirements. (It should be recalled that both Team Corp. and Neotrad Corp. received their equipment at about the same time, some two and a half years prior to my visit, and the study by Blumberg and his colleagues was also conducted some five years after installation; so these results did not merely reflect ramp-up conditions.) Both Blum Corp. and Team Corp. produced a small number of very complex parts, with correspondingly more difficult tasks in machining and inspection. Moreover, as mentioned earlier, in both these installations, procurement of sufficiently high-quality castings was difficult. Neotrad Corp. produced a large number of much simpler parts. Neotrad Corp.'s scheduling and proofing challenge was greater, but once those programs were established, the system suffered fewer interruptions.

DISCUSSION

In this section I confront the comparative case descriptions with the expectations derived from the background literature and suggest some issues for future research.

Automation and work requirements

The data on Team Corp. and Neotrad Corp. suggest that FMS, whether implemented in team form or in more conventional ways, have higher work requirements in most dimensions than either form of stand-alone technology. These data thus suggest that the Braverman de-skilling hypothesis does not fit the reality of these two facilities.

While a team work-design reinforced this general upgrading, it was not required to turn a de-skilling tendency into upgrading. In the cases studied, different management work-design strategies do not seem to have greatly influenced the assessment of work requirements.

Former NC operators did not experience a greater increase in work requirements than former conventional machinists, so the curvilinear hypothesis is not supported either. By the same token, however, the data do not support the hypothesis that NC operators are more skilled than conventional machinists. In their old jobs, these former NC operators may have needed a little less experience and may have felt greater boredom, but the evidence suggests that they needed about the same level of skill, responsibility, teamwork, training and concentration as their conventional machinist colleagues.

Given these results, future research could profitably attempt to specify more precisely what *types* of skills, experience, sense of responsibility and so forth are needed for effective FMS operations. Human Resource policies need this more precise characterization in order to determine optimal selection, training, and industrial relations policies.

Job characteristics and work outcomes

My comparison of the three FMS installations suggests, first, that Blum Corp. might not be very representative of FMS installations. Several contextual and operating conditions combined to make Blum Corp. workers particularly and understandably dissatisfied with important aspects of their jobs.

Second, the rankings of the job characteristics model variables show results somewhat at variance with the expectations drawn from the job-design literature: Team Corp., despite its innovative work design, did not seem to distinguish itself in its job characteristics, and if any system evidenced superior worker reactions and performance, it was Neotrad Corp.'s.

These results suggest several issues for future research. First, more research on teamwork job design might be useful. Blumberg and Gerwin conclude their study of FMS with a policy recommendation: 'Where work structure is based upon traditional job classification and hierarchical leadership it is likely that direct workers will suffer from a lack of control. One alternative to consider is a semi-autonomous work group' (1984: 127–8). The replication of the analysis that led them to this conclusion, and in particular the analysis of an FMS with a team philosophy, suggests that team organization may not be necessary to avoid dissatisfaction and inefficiency on FMS installations. A more traditional job design can achieve high levels of worker motivation and productivity – if, like Neotrad Corp. but unlike Blum Corp., it is implemented with some flexibility and attention to individual preferences.

Behind this pragmatic issue lies a more theoretical issue that would also merit further research. While we were not able to collect any compelling evidence to support this proposition, our interviews with several workers at Neotrad Corp. and Team Corp. suggested that job specialization in Neotrad Corp. was not a source of frustration for two inter-related reasons: (a) it was perceived by the workers as an effective way to get the job done, and (b) it was not used as a social control mechanism nor to limit their promotion opportunities. In this sense, my results are interesting to compare to those of Organ and Greene (1981) and Podsakoff, Williams and Todor (1986), who found that formalization and reduction in autonomy are often negatively, not positively associated with alienation. The result is surprising when we consider the long history of sociology's critique of bureaucracy and other forms of standardization and formalization that turn the employee into a mere cog in the system. But the new research suggests that when workers can establish a feeling of organization-wide responsibility for the effectiveness of their work, sacrifices of individual autonomy and even sacrifices of work group autonomy can be accepted as long as these sacrifices are seen as effective ways to accomplish necessarily interdependent and routine tasks. Under these conditions, low individual autonomy and even low work group autonomy can coexist with relatively high satisfaction and motivation. FMS are an interesting case because, in order for the FMS team to be effective, it must relinquish some of its autonomy in favour of a broader network of agents including other shifts, support functions in the plant, and suppliers.

If we push this analysis a step further, I am thus led to hypothesize that it is the notion of autonomy that leads us astray. Autonomy is the absence of external constraint; but the key factor behind motivation and satisfaction might be the obverse, that is efficacy, the power to accomplish significant objectives (see also Sutton and Kahn, 1987). If Neotrad Corp. workers showed a high level of the JDS 'critical psychological states', it is perhaps because their job design – even though not intrinsically very motivating by the Hackman/Oldham criteria – fitted well the nature of the task they were confronted with (Morse and Lorsch, 1970). Perhaps this proposition should be reformulated in more contingent terms: when workers take a purely instrumental attitude to their work, autonomy as absence of constraint may be a good predictor of satisfaction; but if workers identify with the broader goals of their work, autonomy may be less salient than efficacy.

These findings clearly carry potentially very significant implications both for managerial practice and for the future directions which future research might profitably take. Both of these sets of implications are explored in the following, final, chapter.

This chapter draws on 'Workers' Assessments of Three Flexible Manufacturing Systems.' *International Journal of Human Factors in Manufacturing*, Vol. 1, no. 1, Jan. 1991, pp. 35–54, and 'Workers and Flexible Manufacturing Systems: Three Installations Compared.' *Journal of Occupational Behaviour*, Vol. 12, no. 5, September 1991, pp. 447–460.

REFERENCES

Adler, P. S. and Borys, B. (1989) Automation and skill: three generations of research on the NC case, *Polices and Society*, forthcoming.

Bell, D. (1973) *The Coming of Post-Industrial Society*, Basic Books, New York.

Blackburn, P., Coombs, R. and Green, K. (1985) *Technology, Economic Growth and the Labour Process*, St Martin's Press, New York.

Blauner, R. (1964) *Alienation and Freedom: The Factory Worker and his Industry*, University of Chicago Press.

Blumberg, M. (1980) Job switching in autonomous work groups: an exploratory study in a Pennsylvania coal mine. *Academy of Management Journal*, Vol. 23, pp. 287–306.

Blumberg, M. and Alber, A. (1982) The human element: its impact on the productivity of advanced batch manufacturing systems, *Journal of Manufacturing Systems*, Vol. 1, no. 1, March, pp. 21–30.

Blumberg, M. and Gerwin, D. (1984) Coping with advanced manufacturing technology, *Journal of Occupational Behaviour*, Vol. 5, pp. 113–30.

Braverman, H. (1974) *Labor and Monopoly Capital*, Monthly Review Press, New York.

Child, J. (1987) Organizational design for advanced manufacturing technology, in T. D. Wall, C. W. Clegg and N. J. Kemp (eds.) *The Human Side of Advanced Manufacturing Technology*, John Wiley, Chichester, pp. 101–34.

Cummings, T. and Blumberg, M. (1987) Advanced manufacturing technology and work design, in T. D. Wall, C. W. Clegg and N. J. Kemp (eds.) *The Human Side of Advanced Manufacturing Technology*, John Wiley, Chichester, pp. 37–60.

Ebel, K.–H. (1985) Social and labor implications of flexible manufacturing systems, *International Labour Review*, Vol. 124, no. 2, March–April, pp. 133–45.
Edquist, C. and Jacobsson, S. (1988) *Flexible Automation: The Diffusion of New Technology in the Engineering Industry*, Basil Blackwell, Oxford.
Emery, F. E. (1972) Some hypotheses about the way in which tasks may be more effectively put together to make jobs, in P. Hill (ed.) *Toward a New Philosophy of Management*, Appendix 2, Harper & Row, New York.
Goodman, P. S. (1986) Impact of task and technology on group performance, in P. S. Goodman and associates, *Designing Effective Work Groups*, Jossey-Bass, San Francisco.
Graham, M. B. W. and Rosenthal, S. R. (1985) Flexible manufacturing systems require flexible people. Manufacturing Roundtable Research Report, Boston University.
Hackman, J. R. and Oldham, G. R. (1980) *Work Redesign*, Addison-Wesley, Reading, MA.
Hayes, R. H. and Wheelwright, S. C. (1979) Link manufacturing process and product life cycles, *Harvard Business Review*, Jan–Feb, pp. 71–89.
Hickson, D. J., Pugh, D. S. and Pheysey, D. C. (1969) Operations technology and organizational structure: an empirical reappraisal, *Administrative Science Quarterly*, Vol. 14, pp. 378–97.
Jaikumar, R. (1986) Post-industrial manufacturing, *Harvard Business Review*, Nov–Dec, pp. 69–76.
Jones, B. (1985) Flexible technologies and inflexible jobs: impossible dreams and missed opportunities. Society of Manufacturing Engineers, Technical Paper, MM85–728.
Jones, B. and Scott, P. (1986) 'Working the system': A comparison of the management of work roles in American and British Flexible Manufacturing Systems. Paper for the annual conference of the Operations Management Association of Great Britain, University of Warwick, January.
Kelley, M. R. (1987) *Do Blue-Collar Workers Program Computer-Controlled Machines?* University of Massachusetts, Boston.
Kelly, J. E. (1985) Management's redesign of work: labour process, labour markets and product markets, in D. Knights, H. Willmott and D. Collinson (eds) *Job Redesign: Critical Perspectives on the Labour Process*, Gower, London.
Kerr, C., Dunlop, J. T., Harbison, C. and Myers, C. A. (1964) *Industrialism and Industrial Man*, Oxford University Press, New York.
Kiggundu, M. N. (1981) Task interdependence and the theory of job design, *Academy of Management Review*, Vol. 6, no. 3, pp. 499–508.
Kohler, C. and Schultz-Wild, R. (1985) Flexible manufacturing systems: manpower problems and policies, *Journal of Manufacturing Systems*, Vol. 4, no. 2, June, pp. 101–115.
Manufacturing Studies Board (1986) *Human Resource Practices for Implementing Advanced Manufacturing Technology*, National Academy Press, Washington DC.
Morse, J. J. and Lorsch, J. W. (1970) Beyond theory Y, *Harvard Business Review*, May–June, pp. 61–8.
Oldham, G. R., Hackman, J. R. and Stepima, L. P. (1979) Norms for the job diagnostic survey, *JSAS Catalog of Selected Documents in Psychology*, Vol. 9, no. 14, (Ms. 1819).

Organ, D. W. and Greene, C. N. (1981) The effects of formalization on professional involvement: a compensatory process approach, *Administrative Science Quarterly*, Vol. 26, pp. 237–52.

Pasmore, W., Francis, C. and Haldeman, J. (1982) Sociotechnical systems: a North American reflection on empirical studies of the seventies, *Human Relations*, Vol. 35, no. 12, pp. 1,179–204.

Piore, M. J. and Sabel, C. F. (1984) *The Second Industrial Divide: Possibilities for Prosperity*, Basic Books, New York.

Podsakoff, P. M., Williams, L. J. and Todor, W. D. (1986) Effects of organizational formalization on alienation among professionals and nonprofessionals, *Academy of Management Journal*, Vol. 29, no. 4, pp. 820–31.

Quinn, R. P. and Shepard, L. J. (1974) *The 1972–1973 Quality of Employment Survey*. Survey Research Center, University of Michigan, Ann Arbor.

Quinn, R. P. and Staines, G. L. (1979) *The 1977 Quality of Employment Survey*. Survey Research Center, University of Michigan, Ann Arbor.

Rousseau, D. M. (1977) Technological differences in job characteristics, employee satisfaction, and motivation: a synthesis of job design research and sociotechnical systems theory, *Organizational Behavior and Human Performance*, Vol. 19, pp. 18–42.

Schultz-Wild, R. and Kohler, C. (1985) Introducing new manufacturing technology: manpower problems and policies, *Human Systems Management*, Vol. 5, pp. 231–43.

Seppala, P., Tuominen, E. and Koskinen, P. (1985) Job structure and work content in a flexible manufacturing system: analysis of a case from the Finnish engineering industry, in P. Brodner (ed.) *Skill Based Automated Manufacturing*, Pergamon Press, Oxford.

Spenner, K. I. (1983) Deciphering Prometheus: temporal change in the skill level of work, *American Sociological Review*, Vol. 48, pp. 824–37.

Steffy, W., Smith, D. N. and Souter, D. (1973) *Economic Guidelines for Justifying Capital Purchases*, Industrial Development Division, Institute of Science and Technology, University of Michigan, Ann Arbor.

Sutton, R. I. and Kahn, R. L. (1987) Prediction, understanding and control as antidotes to organizational stress, in J. W. Lorsch (ed.) *Handbook of Organizational Behavior*, Prentice-Hall, Engelwood Cliffs, NJ, pp. 272–85.

Thompson, J. D. (1967) *Organizations in Action*, McGraw Hill, New York.

Toikka, K. (1985) Development of work in FMS – case study on new manpower strategy, in P. Brodner (ed.) *Skill Based Automated Manufacturing*, Pergamon Press, Oxford.

Walton, R. E. (1977) Improving the quality of work life, *Harvard Business Review*, May–June.

Walton, R. E. and Susman, G. I. (1987) People policies for the new machines, *Harvard Business Review*, March–April, pp. 77–84.

Woodward, J. (1958) *Management and Technology*, HMSO, London.

12

FUTURE PROSPECTS

John Storey

The issues examined in this book will undoubtedly continue to be debated by managers, policy-makers, academics and others for a long time to come. The discussion so far and the current level of knowledge provoke three fundamental questions for consideration. First, what are the characteristics of the New Wave Manufacturing methods which seem to be responsible for superior competitive performance? A key aspect of this question, as we have seen in this book, is the extent to which technical and/or organizational characteristics are critical. Second, to what degree is the 'winning formula' amenable to amendment and adjustment? In particular here, a central issue is whether the unwelcome features of Japanese manufacturing methods can be avoided. These issues carry important implications for policy-makers and practitioners. Thirdly, and finally, how, in the future, might these questions best be researched? This concluding chapter is structured around these three sets of questions.

THE CENTRAL CHARACTERISTICS OF NEW WAVE MANUFACTURING

Even when one takes into account the hype, the commercial promotion, the exaggeration and the myths, the fact remains that certain manufacturing facilities are patently far more productive, cost competitive and reliable than the rest. Moreover, one can go further and observe that this superiority in production (not necessarily in social terms) is demonstrated even without any marked technological advantages relating, for example, to automation.

Such startling results inevitably prompt a search for the *essential* (as opposed to the peripheral) characteristics of this 'superior form'. These features have been explored throughout this book in some detail. They include: a fuller utilization of available work time; flexibility of work and of labour deployment; teamworking of one kind or another; Just-in-Time production; continued improvement, learning by doing and innovative ideas contributed by all levels of employees; the continual bearing down, and elimination of, non-value-added activities; and workers undertaking production, inspection and maintenance functions themselves.

These and similar features have been identified in numerous studies. There is not, in fact, as much controversy as one might expect concerning the content of this kind of list: the real controversy concerns the relative importance of different items on the list, the explanation for the origins of these ways of working and their sets of consequences – particularly for employees.

ADOPTING AND ADAPTING THE NEW MANUFACTURING

Problems and dilemmas arise when attempts are made to isolate the crucial factors which are perceived as making the *real* difference. For, despite the widely cited 'package' of features, different commentators tend to point to certain particular elements as being rather more fundamental than others in explaining the productivity performance gap. Depending upon which form of explanation one favours, the implications for practitioner choice and action are extensive.

Let us take each of the main explanations in turn (each carries its own implicit and even, occasionally, explicit implications for practical action). First, there is the view that the main competitive weapon is Advanced Manufacturing Technology (AMT). According to this view the lesson is that Western manufacturing companies simply must keep pace with leading Japanese and other competitors in terms of investment in AMT and in the appropriate utilization of AMT. There is some research evidence, for example, which reveals that Western countries have fallen behind Japan in the adoption of Flexible Manufacturing Systems (FMS) (Jaikumar, 1986; Valery, 1987). The Japanese had more than twice the number of FMS of the USA, for example. Moreover, American corporations were found to be not utilizing FMS properly – they were failing to take advantage of its potential. For instance, the average number of component parts produced per FMS in the USA was ten, whereas in Japan it was ninety-three (Kenney and Florida, 1988: 141).

But, this kind of 'technological explanation' has been severely challenged by many seasoned practitioners and academic researchers who have visited Japanese companies. Indeed, one of the most consistent messages emanating from the study tours is that the competitive edge of Japanese manufacturers does *not* derive from technological superiority. In fact, in many areas of AMT and automation, the Japanese plants could be said to 'lag' behind the Americans – and yet still enjoy a competitive advantage as measured by productivity. Hence, analysts have been forced to dig a little deeper. The result has been that instead of primarily emphasizing further investment in AMT as the answer, analysts have largely concurred that the advantage gained through the new manufacturing methods at Toyota and elsewhere results from *the social organization of production.*

This form of explanation has two different branches. The first stresses the Human Resource Management aspects such as employee participation, teamworking, security of employment, commitment and extensive training. The implications for practice which stem from this type of explanation are relatively clear. In order to compete under these terms a Human Resource Management strategy would need to be devised and Human Resources, overall, elevated in importance. The model would point, for example, to a company extricating itself from conventional Tayloristic and Fordist principles which stress a high division of labour, detailed job specification and minimum scope for the exercise of employee discretion, short job cycles, and a basic mistrust of shopfloor employees. Guides to action for those choosing to go down this Human Resource Management route can be readily found (e.g. see Dale and Cooper, 1992; Walker, 1992; Armstrong, 1992).

But analysts who take the second branch of explanation concerning the social organization of production would tend to counsel caution here. The real source of competitive advantage enjoyed by new wave manufacturers such as Toyota stems, they say, not from Human Resource policies and practices but from production management arrangements. This type of explanation is, for example, favoured by Schonberger (1982). According to this line of reasoning it is simply mistaken to interpret the leading Japanese manufacturers as having abandoned Taylorism. On the contrary, work study, industrial engineering, standardized and repetitious work patterns, job simplification, the establishment of time and production standards and, above all, extensive managerial control, are all flourishing in Japan. Indeed, these features are taken to new heights because employee groups are themselves engaged in seeking out 'unnecessary movements' and excess labour in true Tayloristic fashion.

Underpinning systematic production control methods are a range of work intensification devices which ensure that productivity measures are high because labour content is continually being taken out. Schonberger (1982: 91) points to the management practice of 'deliberately pulling workers off the line when the line is running smoothly'. Drawing on his studies at Kawasaki, he describes how assemblers are instructed to turn on a yellow light above their workstation when they cannot keep up. This summons foremen and other workers to help. Interestingly, Schonberger found that Kawasaki managers are *pleased* 'when many yellow lights are on' because, if no yellow lights are on 'management knows that the line is moving too slowly or there are too many workers'.

The implications for management practitioners deriving from this production control thesis point to a rather different set of practical measures from those advocated above. For a start, the production management explanation of Japanese success would seem to point to the retention rather than the abandonment of Taylorism. The practical feasibility of this sort of programme might be called into severe question in the West. Indeed, the explanation for employee compliance with this level of work intensification in Japan itself has given rise to controversy. Many commentators point to the cultural differences and emphasize 'groupism' in Japan, while noting the 'late-developer effect', the 'feudalistic legacy' and the profound impact of defeat in the Second World War. A severely amended version of the culture thesis is advanced by Dohse, Jurgens and Malsch (1986). They suggest that the source of productivity advantage does largely arise from a management system which intensifies work, but that in order to explain the viability of this in the Japanese context, one must take account of the assertion of managerial prerogatives which occurred through decisive defeats inflicted upon organized labour in the 1950s.

The practical implications of this line of reasoning for people in the West would seem to point in two different directions. Either it is maintained that such an assertion of managerial prerogatives would be neither feasible nor desirable in the Western economies, or, one might argue that developments in recent years in the USA and Europe indicate just such an unfolding of events.

THE FUTURE RESEARCH AGENDA

There are evidently many questions which require further investigation. The 'long list' of features associated with the new model way of manufacturing used by the most competitive producers can be constructed with some confidence. What is far less certain is the 'short list' of core, irreducible, elements that arguably constitute the real cutting edge. Is the heart of the matter, for example, as Wickens (1993) suggests, 'people care', continuous development, flexibility and teamworking? Or, was Schonberger nearer the mark when he emphasized the production control elements? Alternatively, it might be that a third position is more tenable – one which attributes success in manufacturing to what Shimada (1993: 27) describes as the 'intersection' between technology and social organization, that is 'humanware'. Shimada's model implies a close integration between certain technological features and certain principles of social organization. In a simplified format this can be depicted as in Figure 12.1.

The model is built around various hypothesized inter-connections. These ought to be amenable to empirical testing. For example, the critical connections relate to the sources of 'adaptability' and the idea of 'human control'. The bottom half of the figure seems rather less contentious but the hypothesized causal relationships in the top half of the figure would seem to deserve further investigation.

The construction of hypotheses for testing should be relatively straightforward. The importance of including the multiple features of technology, production control, industrial relations, and Human Resource Management practices has already been made clear. Another part of the research effort must also be directed to the study of worker responses to New Wave Manufacturing of the kind discussed by Adler in Chapter 11. Adler demonstrated a controlled way in which comparative research can be conducted. Further work is also required on the particular skill profiles and behaviour patterns which are at a premium under the new conditions. Following on from this, what Human Resource policies and practices seem most conducive to the promotion of these behaviour characteristics and competences?

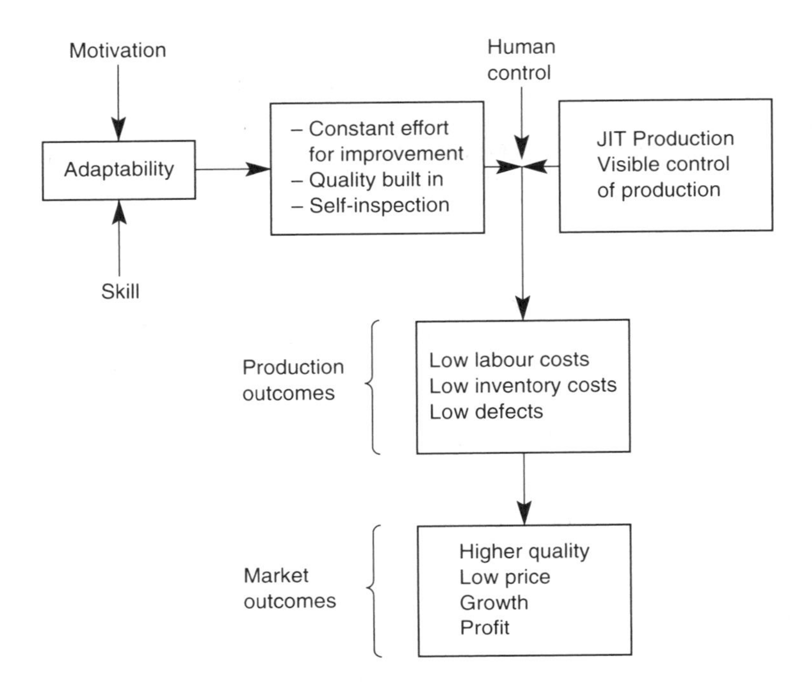

Figure 12.1 A simplified (amended) version of Haruo Shimada's 'humanware' model (Based loosely on Shimada, 1993: 28)

A key lesson of this book, however, is that not all of the research endeavour should be spent on the 'shopfloor' issues of technology, people and work systems – vital though the issues at this level certainly are. As the earlier chapters in the book demonstrated, there are important issues deserving of study at the strategic level. Chapters 2 and 3, for example, made some tentative links between business strategies of cost leadership, quality or innovation and the particular characteristics of different new manufacturing methods. This line of reasoning is ripe for further research. What type of business or marketing strategy is, for example, made possible through the adoption of JIT or MRPII? Do firms (*should* firms) think through their market positioning before investing in one of these systems or is the rather more generalized drive for 'improvement' sufficient – or perhaps even preferable? The implication arising from Chapter 4 is that, in fact, incremental change will not be enough: the thrust of Francis's argument is that a throughgoing and well-thought-through step-change is required. Even this, he suggests, may not be enough: new institutions above and beyond the level of the firm may need to be created. The research task is demanding indeed!

Problematical also is the question of precisely *how* these issues can be researched. What research methods would be most appropriate? More detailed and longer-term case studies of Japanese companies which are at the forefront of the new methods can certainly be recommended. So too, must studies continue of course in those Western corporations using, or seeking to introduce, new manufacturing methods. Ideally, some of these studies will use action research methods where incremental change can be closely monitored and evaluated. In addition, further careful studies of the *pattern* of organizational arrangements and managerial practices associated with the successful utilization of AMT can also be recommended (for a useful example of this kind of research see Bessant, 1993). But the logic of Francis's analysis in Chapter 4 is that the analysis must also extend beyond the individual plant and company.

As the chapters in this book will have surely revealed, the research agenda is very extensive. The issues at stake are fundamental – industrial competitiveness, managerial control, employee commitment, the nature and design of work. They are issues which truly lie at the heart of social science. The practical consequences are enormous – standards of living and the experience of work to name just two. It is to be hoped that future research will be able to do justice to the enormity of this challenge.

REFERENCES

Armstrong, M. (ed.) (1992) *Strategies for Human Resource Management: A Total Business Approach*, Kogan Page, London.

Bessant, J. (1993) Towards Factory 2000: designing organizations for computer-integrated technologies, in J. Clark (ed.) *Human Resource Management and Technical Change*, Sage, London.

Dale, B. G. and Cooper, C. (1992) *Total Quality and Human Resources*, Blackwell, Oxford.

Dohse, K., Jurgens, U. and Malsch, T. (1986) From Fordism to Toyotism? The social organization of the labour process in the Japanese automobile industry, *Politics and Society*, Vol. 14, pp. 115–46.

Jaikumar, R. (1986) Post industrial manufacturing, *Harvard Business Review*, November–December, pp. 69–76.

Kenney, M. and Florida, R. (1988) Beyond mass production: production and the labour process in Japan, *Politics and Society*, Vol. 16, no. 1, pp. 121–58.

Schonberger, R. J. (1982) *Japanese Manufacturing Techniques*, Free Press, New York.

Shimada, H. (1993) Japanese management of auto production in the United States: an overview of 'Human Technology', in ILO, *Lean Production and Beyond*, International Labour Office, Geneva.

Valery, N. (1987) Factory of the future, *The Economist*, 30 May, pp. 1–18.

Walker, J. W. (1992) *Human Resource Strategy*, McGraw Hill, New York.

Wickens, P. (1993) Lean, people-centred, mass production, in ILO, *Lean Production and Beyond*, International Labour Organization, Geneva.

INDEX